Philip Atkinson

The Electric Transformation of Power and its Application by the Electric Motor

Philip Atkinson

The Electric Transformation of Power and its Application by the Electric Motor

ISBN/EAN: 9783744649841

Printed in Europe, USA, Canada, Australia, Japan

Cover: Foto ©berggeist007 / pixelio.de

More available books at **www.hansebooks.com**

THE ELECTRIC TRANSFORMATION OF POWER

AND ITS APPLICATION BY

THE ELECTRIC MOTOR,

INCLUDING

ELECTRIC RAILWAY CONSTRUCTION.

BY

PHILIP ATKINSON, A.M., PH.D.,

Author of "Elements of Static Electricity,"
"The Elements of Electric Lighting,"
"The Elements of Dynamic Electricity and Magnetism."

NEW YORK:
D. VAN NOSTRAND COMPANY
23 MURRAY AND 27 WARREN STREET.

LONDON:
CROSBY LOCKWOOD AND SON,
7, STATIONERS' HALL COURT, LUDGATE HILL.
1893.

INTRODUCTION.

THE design of this book is to give, in plain, untechnical language, the essential facts in regard to the means by which electricity is employed as an agent for the transformation and transmission of power, and its application to the operation of machinery. These facts comprehend the construction and principles of the electric motor, and its relations to the dynamo, and through it to the steam engine, water-wheel, or other source of power.

The selection of motors as examples of the different kinds of construction has been made from those which have stood the test of practical use, and embody the leading principles of the best construction; the selections being necessarily limited to a few of the many excellent motors now in use.

Various special applications of the motor have been given, to show its adaptability to the different kinds of mechanical work, and to the difficult conditions often met with in its performance. And the construction and operation of the electric railway, including its motors, cars, and various auxiliary apparatus, have received the full consideration which the importance of the subject demands.

All historical matter has been excluded, and the description of all experimental apparatus, and of apparatus which, though still in use, has been superseded; only the latest and most approved kinds, which have had more or less practical development, being described.

Brief definitions of such technical terms as are indispensable to a proper understanding of the subject have been given in the first chapter, which is intended chiefly for reference; the values of the electric units being given in absolute measure, based on the fundamental C. G. S. units, from which an accurate estimate of the quantity of power which any motor or dynamo is designed to transform can be obtained, as indicated by its electric energy when given in such units.

The thanks of the writer are due to the various electric companies whose apparatus has been described, for valuable technical information and the use of cuts; also to the electric journals, especially The Electrical Engineer, for similar favors; and to friends who have kindly furnished such information on various points.

PHILIP ATKINSON.

CHICAGO, August 17, 1893.

CONTENTS.

CHAPTER I.

DEFINITIONS 1   PAGE

    The Conservation of Energy. Electric Terms. Potential. Electromotive Force. Resistance. Current. Ohm's Law. Induction. C. G. S. Units. The Dyne. The Erg. Electric Units. The Volt. The Ohm. The Ampere. The Watt. The Electric Horse-Power. Table of Dimensions and Resistances of Pure Copper Wire.

CHAPTER II.

PRINCIPLES OF THE ELECTRIC MOTOR 7

    The Motor a Means of Applying Power. Sources of Electric Energy. Classification of Motors. Construction of Direct Current Motors. The Armature. The Commutator. The Brushes. The Field-Magnets. The Series Wound Motor. The Shunt Wound Motor. Operation of the Motor as a Dynamo. Commutation. Polarity and Neutral Line. Coreless Construction. Reversal of Rotation. Position of the Brushes. Operation of the Direct Current Motor. Effect of Current Reversal on Rotation. Polar Rotation. Counter Electromotive Force. Position of the Poles and Neutral Line. Eddy Currents. Loss of Power. Series and Shunt Motors Compared. Constant Current and Constant Potential Circuits and Motors. The Rheostat. Rheostat Connections. Electric Heat and Mechanical Energy. Motor Designing.

CHAPTER III.

STATIONARY MOTORS 42

The Excelsior Motor. The Edison Standard Motor. The Edison Small Motor. The C. & C. Standard Motor. The C. & C. Small Motors. The Detroit Motor. The Eddy Motor. The Perret Motor. Alternating Current Motors. The Tesla Alternating Current Motor. The Stanley-Kelly Alternating Current Motor. Single Phase Alternating Current Motors. The Brown Single Phase Alternating Current Motor.

CHAPTER IV.

APPLICATIONS OF THE STATIONARY MOTOR 81

General Remarks. Electric Fans and Ventilators. Electric Operation of Pipe Organs. Electric Elevators. The Otis Electric Elevator. The Sprague-Pratt Electric Elevator. Electric Dock Hoists. Electric Travelling Cranes. Electric Operation of Printing Presses. Commercial Measurement of Electric Energy. Electric Operation of Dental Apparatus. Electric Operation of Medical and Surgical Apparatus. Electric Operation of Ship Drills. The Edison Electric Percussion Drill. The Van Depoele Electric Percussion Drill. The Electric Diamond Drill. The Triplex Electric Pump. The Sperry Pick Electric Coal Cutter. The New Arc Electric Coal Cutter. Various Electric Mining Apparatus.

CHAPTER V.

ELECTRIC RAILWAYS AND RAILWAY MOTORS 137

General Remarks. Line Construction. Feeders. Poles. Trolleys. The Boston Trolley. The Emmet Trolley. The Compression Spring Trolley. The Siemens-Halske Sliding Contact. The Tube and Piston Contact. The Double Trolley. Insulators and Clamps. Switches. Three-way Switch. The Emmet Switch. The Atkinson Switch. Right-Angled Crossing. The Ramsay Adjustable Crossing. Trolley Line-Breaker. The Johnston Disconnector. Railway Motors. The Westinghouse Single Reduction Motor. Controller. Controller Connections. Lightning-Arrester. The Thomson-Houston Water-Proof Single Reduction Motor. The Curtis

CONTENTS.   vii

                                                              PAGE
Single Reduction Motor. Gearless Motors. The Short Gearless Motor. Electric Lighting of the Cars. Electric Heating of the Cars. The Burton Electric Heater. The Conduit System. The Siemens-Halske Conduit Railway. The Love Conduit Railway. Closed Conduits. The Wheless Conduit Railway. Elevated and Underground Electric Railways. The Liverpool Elevated Electric Railway. The City and South London Underground Electric Railway. Storage Battery Traction. Electric Haulage in Mines. Electric Haulage in Mills and Factories.

CHAPTER VI.

CENTRAL STATION CONSTRUCTION AND EQUIPMENT 208
    Development of the Central Station. Chicago Edison Central Station No. 1. Chicago Edison Central Station No. 5. Cicero and Proviso Street Railway Central Station. Waterpower Stations and Long Distance Transmission. The Frankfort-Lauffen Experiment. Willamette Falls Waterpower Station. Telluride Waterpower Station. Direct Connected Dynamos.

THE ELECTRIC TRANSFORMATION OF POWER.

CHAPTER I.

DEFINITIONS.

The Conservation of Energy.—As the conservation of energy is the cardinal principle which underlies the electric transformation of power, a brief explanation of this doctrine is appropriate as an introduction to our subject, and will aid materially in its proper elucidation.

Energy is the inherent power by which matter operates, or performs work, and without which it would be wholly inert. Like matter it exists under a great variety of forms, and may be transmuted, or changed from one form to another, but can neither be created nor destroyed. It may disappear in one form and reappear in another, or in various forms, but in all these transmutations there is no loss, the energy which reappears being mathematically equal in quantity to that which disappeared.

Among the various forms of energy may be mentioned vital force, chemical affinity, molecular repulsion, gravity, heat, light, electricity, magnetism, and mechanical power.

As an instance of transformation, the energy of gravity as developed in a water-fall may be selected. This may be converted into mechanical power by a water-wheel, this power converted into electricity by a dynamo, and developed in part as light and heat in an electric lamp, and in part as mechanical power by an electric motor. In each of these transformations a certain percentage of the energy is consumed either in overcoming friction and inertia or electric resistance, while another portion is dissipated into the air and surrounding objects. But if all the energy thus consumed and dissipated in transit could be collected and added to that finally developed as light, heat, and mechanical power it would be found exactly equal in amount to that of the gravity expended to produce it; a proposition which has been fully demonstrated by carefully conducted laboratory experiments, in which the energy expended has been reproduced, and whose truth is practically apparent in all mechanical operations.

Electric Terms.—The principal electric terms employed in this book may be briefly defined as follows:

Potential.—Electric potential is that condition in bodies which represents the relative electric energy manifested by them. When the quantity of this energy differs in bodies associated together, those in which it is greater are said to have positive potential with respect to those in which it is less, and the latter to have negative potential with respect to the former; but when it is equal in all, they are said to have zero potential with respect to each other.

Electromotive Force.—Electromotive force is that electric condition which results from difference of potential between bodies associated together by which electric energy is transmitted from those having positive potential to those having negative; hence it is often represented as *electric pressure*. It may exist in the earth or atmosphere as the result of natural causes, or be produced by electric generators, as the dynamo or battery, when required for practical work. Its symbol is E. M. F. or E alone.

Resistance.—Electric resistance, symbol R, is that which opposes the transmission of electric energy in bodies. It consists chiefly either in the molecular constitution of the body, or in counter electromotive force. Molecular resistance is found in all bodies; those in which it is relatively small being known as *conductors*, and those in which it is so large as practically to suppress electric transmission, being known as *non-conductors* or *insulators*. Conductors may have high or low resistance, and insulators, high or low insulation. This resistance varies directly as the length of the body and inversely as its cross-section. Counter electromotive force may be produced by the direct action of an opposing generator, or indirectly by the inductive influence of electric transmission through a parallel conductor.

Current.—Electric current, symbol C, is that electric condition which results from the transmission of electric energy through a conductor, by electromotive force, in opposition to electric resistance.

Ohm's Law.—According to the law discovered by Ohm, *the strength of an electric current varies directly as the electromotive force by which the current is impelled, and inversely as the total resistance encountered.* Which may be expressed by the formula $C = \dfrac{E}{R}$, from which the values of E and R may be derived as follows, $E = CR$, $R = \dfrac{E}{C}$. That is, the current strength equals the electromotive force divided by the resistance; the electromotive force equals the current strength multiplied by the resistance; the resistance equals the electromotive force divided by the current strength.

Induction.—Electric induction is the influence which an electrified body, as a current-bearing wire, or a body bearing a static charge, exerts on other bodies in its vicinity, from which it is insulated. Magnetic induction is analogous in character, and pertains to bodies capable of exerting or receiving magnetic influence.

C. G. S. Units.—Certain established units are required by which electric energy may be calculated; these are based on the centimeter-gramme-second (C. G. S.) system of units, in which the *centimeter* is the unit of *length*, the *gramme* the unit of *mass*, and the *second* the unit of *time*. Two of the principal units in this system are the *dyne*, the unit of *force*, and the *erg*, the unit of *work*.

The Dyne.—The dyne represents the force which, in 1 second, can impart to a mass of 1 gramme a velocity of 1 centimeter per second.

The Erg.—The erg represents the work performed in moving a body through a distance of 1 centimeter with a force of 1 dyne.

Electric Units.—The principal electric units based on this system and referred to in this book are as follows:

The Volt.—The unit of electromotive force, or electric pressure, is the volt, a definite idea of the value of which may be obtained from the statement, that it requires 100,000,000 ergs of work to cause 1 C. G. S. unit of electricity to flow past a point, when urged by an electric pressure of 1 volt. Hence the volt represents 100,000,000 electromagnetic C. G. S. units of electromotive force, which is nearly equal to the electromotive force of a Daniell battery cell, from which the value of this unit was derived; and this number is more conveniently expressed by the form 10^8.

The Ohm.—The ohm is the unit of electric resistance, and represents $1,000,000,000 = 10^9$ electromagnetic C. G. S. units of electric resistance; which equals the electric resistance of a column of pure mercury $106\frac{21}{100}$ centimeters in length, and 1 square millimeter in cross-section, at the temperature of zero centigrade, or freezing point.

The Ampere.—The ampere is the unit of electric current, and represents such a current as would traverse an electric circuit whose resistance is 1 ohm, when urged by an electromotive force of 1 volt; and is obtained, according to Ohm's law, by dividing the value of the volt by that of the ohm. Hence, since $\frac{E}{R} = C$, $\frac{10^8}{10^9} = 10^{-1} = \frac{1}{10}$. The ampere therefore represents $\frac{1}{10}$ of an electromagnetic C. G. S. unit of electric current.

The Watt.—The energy by which electric work is performed includes both electromotive force and electric current, and must therefore be expressed by their joint product, which is represented by the unit known as the watt, obtained by multiplying together the values of the volt and the ampere, thus, $10^8 \times 10^{-1} = 10^7$. The watt therefore represents 10^7 electromagnetic C. G. S. units of electric energy, or ergs per second.

The Electric Horse-Power.—The electric equivalent of the mechanical horse-power is known as the electric horse-power, h. p., and is represented by 746 watts, and hence equal to $746 \times 10^7 = 7,460,000,000$ electromagnetic C. G. S. units of electric energy, or ergs per second.

TABLE OF DIMENSIONS AND RESISTANCES OF PURE COPPER WIRE.

American Gauge. Brown & Sharpe's Number.	Diameter, in Mils. 1 Mil. = .001 in.	Area in Cross-section. Circular Mils. Diameter².	Weight and Length.		Resistance @ 75° F.	
			Lbs. per 1000 ft.	Feet per lb.	Ohms per 1000 feet.	Feet per Ohm.
0000	460.000	211600.00	639.33	1.56	.04906	20383.
000	409.640	167805.00	507.01	1.97	.06186	16165.
00	364.800	133079.40	402.09	2.49	.07801	12820.
0	324.950	105592.50	319.04	3.13	.09831	10409.
1	289.300	83694.20	252.88	3.95	.12404	8062.3
2	257.630	66373.00	200.54	4.99	.15640	6393.7
3	229.420	52634.00	159.03	6.29	.19723	5070.2
4	204.310	41742.00	126.12	7.93	.24869	4021.0
5	181.940	33102.00	100.01	10.00	.31361	3188.7
6	162.020	26250.50	79.32	12.61	.39546	2528.7
7	144.280	20816.00	62.90	15.90	.49871	2005.2
8	128.490	16509.00	49.88	20.05	.62881	1590.3
9	114.430	13504.00	39.56	25.28	.79281	1261.3
10	101.890	10381.00	31.37	31.38	1.	1000.0
11	90.742	8234.00	24.88	40.20	1.2607	793.18
12	80.808	6529.90	19.73	50.69	1.5898	629.02
13	71.961	5178.40	15.65	63.91	2.0047	498.83
14	64.084	4106.80	12.41	80.59	2.5908	385.97
15	57.068	3256.7	9.84	101.63	3.1150	321.02
16	50.820	2582.9	7.81	128.14	4.0191	248.81
17	45.257	2048.2	6.19	161.59	5.0683	197.30
18	40.303	1624.3	4.91	203.76	6.3911	156.47
19	35.390	1252.4	3.78	264.26	8.2889	120.64
20	31.961	1021.5	3.09	324.00	10.163	98.401
21	28.462	810.10	2.45	408.56	12.815	78.037
22	25.347	642.70	1.94	515.15	16.152	61.911
23	22.571	509.45	1.54	649.66	20.377	49.087
24	20.100	404.01	1.22	819.21	25.695	38.918
25	17.900	320.40	.97	1032.96	32.400	30.864
26	15.940	254.01	.77	1302.61	40.868	24.469
27	14.195	201.50	.61	1642.55	51.519	19.410
28	12.641	159.79	.48	2071.22	64.966	15.393
29	11.257	126.72	.38	2611.82	81.921	12.207
30	10.025	100.5	.30	3293.97	103.30	9.6812
31	8.928	79.71	.24	4152.22	127.27	7.8573
32	7.950	63.20	.19	5236.66	164.26	6.0880
33	7.080	50.13	.15	6602.71	207.08	4.8290
34	6.304	39.74	.12	8328.30	261.23	3.8281
35	5.614	31.52	.10	10501.35	329.35	3.0363
36	5.000	25.00	.08	13238.83	415.24	2.4082
37	4.453	19.83	.06	16691.06	523.76	1.9093
38	3.965	15.72	.05	20854.65	660.37	1.5143
39	3.531	12.47	.04	26302.23	832.48	1.2012
40	3.144	9.89	.03	33175.94	1049.7	.9527

CHAPTER II.

PRINCIPLES OF THE ELECTRIC MOTOR.

The Motor a Means of Applying Power.—The electric transformation of power is effected chiefly through the agency of the electric motor, which is simply a means of applying power transmitted by an electric current. This power may originate in the mechanical energy of the steam engine or the water-wheel, and be converted, at its source, into electric energy by the dynamo and transmitted by electric conductors to the place where it is required, and there reconverted into mechanical energy by the motor, and applied to the operation of machinery. Or it may originate in the chemical energy of the primary battery, or be accumulated as chemical energy by the storage battery, when derived as electric energy from the dynamo or primary battery, and be thence transmitted as electric energy to the motor.

Sources of Electric Energy.—The primary battery, as a source of power, is too expensive, inconvenient, and unreliable for ordinary use; and while it has had a limited use in supplying current for the operation of light machinery, it seldom proves satisfactory; and the dynamo has now become the universally recognized agent for the conversion and transmission of power by electric energy, either directly or by the intervention of the storage battery; direct transmission being usually the preferable mode, wherever it is possible to employ it.

Classification of Motors.—Motors may be classified according to their use, as stationary and portable, and according to the nature of the current by which they are operated, as direct current and alternating current. Those operated by the direct current are classified according to their winding, as series wound and shunt wound, and are by far the largest and most practical class; those operated by the alternating current not having yet been made sufficiently practical to come into general use, though their ultimate success, when the construction is more perfectly adapted to the nature of the current seems to be assured by the success already attained. The construction and operation of direct current motors will be considered first.

Construction of Direct Current Motors.—The general principles of construction are the same in the direct current motor as in the direct current dynamo, and hence each may be employed for the same purpose as the other; the motor for the conversion of mechanical energy into electric, and the dynamo for the conversion of electric energy into mechanical. But there are certain specific differences of construction which adapt each to its own special work, to which, in practice, it is confined. The principal parts of each machine are the same, consisting of the armature, the commutator, the brushes, and the field-magnets, each of which may be briefly described as follows:

The Armature.—The armature is a rotating electromagnet, generally of circular shape, mounted on an iron shaft to which it is attached, and from which it is sometimes magnetically insulated. It consists of an iron core, usually laminated to prevent the formation of wasteful eddy currents, being composed of numerous flat rings, disks, or hoops of thin, soft, sheet-iron, insulated from each other by paper, and bolted together; on which are wound coils of copper wire, insulated by cotton wrapping.

Armatures are of two leading types, of which there are numerous modifications, the Gramme, or ring armature, and the Siemens, or drum armature.

In the Gramme armature the coils are wound on a flat ring, either entirely covering the surface of the core, both exterior and interior, or being separated by projections, as in Fig. 1, and the ring being either narrow, as shown, or of any required width.

Each coil consists of several convolutions, and its terminals are connected with adjacent segments of the commutator; two terminals of adjacent coils being connected with the same segment, so that when all the coils are thus connected they form a continuous, closed circuit.

In the Siemens armature the coils are wound on the exterior surface of a cylinder, or drum, as shown in Fig. 2, crossing at the ends, and having their terminals

FIG. 1.

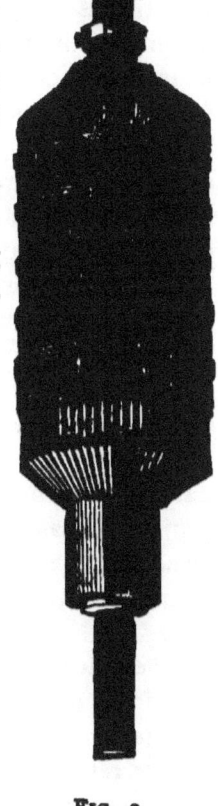

FIG. 2.

connected with opposite segments of the commutator; two terminals of adjacent coils being connected with the same segment on each side, so as to form a closed circuit, as in the Gramme armature. In this armature also the coils are often wound between projections on the core, and the interior ventilated by tubular openings connected with these projections.

Armatures may be constructed with an open circuit, the terminals of each coil being connected with separate segments of the commutator on opposite sides; but such construction is rare, being adopted only in some dynamos, and never now in motors. There are also polar armatures, in which the coils are wound on core projections radiating from a common center; also stationary armatures with rotating field-magnets, and armatures which rotate around stationary field-magnets; but these varieties are all rare.

The Commutator.—The commutator is mounted on the end of the armature shaft, as shown on the left of Fig. 1 and at the bottom of Fig. 2, and consists of a cylinder constructed with copper bars attached to an insulating collar on the shaft, and insulated from each other with mica, usually, so as to furnish a smooth, compact surface; the number of these bars, or segments, being equal to the number of armature coils; two coil terminals, in a closed circuit, being connected with each segment by radial arms or projections, as already described; while, in an open circuit armature, the terminals being connected with separate segments, the number of segments is double the number of coils.

The office of the commutator is to collect the direct current from one terminal of the external circuit, transmit it to the armature, in which it becomes alternating, and thence return it as direct current to the opposite terminal of the external circuit, as explained hereafter.

The Brushes.—The brushes are copper or carbon strips which connect the armature with the field-magnet coils and external circuit through the commutator by contact with its segments. When made of copper they are laminated, each brush being composed of several strips of sheet copper, overlapping each other and sometimes alternating with layers of copper wire, and soldered together at one end, the other end being beveled and maintained in contact with the commutator, in a slanting position, by spring pressure, in such a manner as to bridge the intervening spaces between its segments; the rotation of the commutator being from the inner edge of the bevel toward the point.

The carbon brushes are usually rectangular pieces of carbon, beveled in a similar manner, and sometimes copper-plated to within a short distance of the point to reduce the resistance. They are also made of small cylinders of carbon, three or four in a brush, set in copper sockets. The bearings are often nearly rectangular and the brushes set radially against the commutator, so that rotation in either direction has the same effect on the bearing. As carbon brushes wear the commutator less than copper brushes, they are often preferred, but their high resistance is an objection to their use on motors of low E. M. F.

There must always be two brushes, one on each side of the commutator; but each brush is usually divided into two or three sections, to equalize the pressure and distribute it over the commutator surface, reduce the resistance and heating, and for convenience and economy in replacing worn or broken parts. Two sets of brushes, pointing in opposite directions, are also sometimes employed for reversal of rotation, and the current admitted through either set, according to the direction in which rotation is required; one set being lifted out of contact as the other set is lowered into contact.

The Field-Magnets.—These are stationary electromagnets, usually of the horse-shoe form, having massive iron cores terminating in large pole-pieces, which partly inclose the armature and are placed as close to it as its rotation will permit without contact. The cores are wound with insulated copper wire in such a manner as to produce opposite polarity in the opposite pole-pieces, the winding being either in series or shunt as explained hereafter; compound winding, employed in some of the older motors, not being found so well adapted to internal regulation in motors as in dynamos.

The cores and pole-pieces may be either of cast-iron or wrought-iron, the latter being preferred for both, on account of its higher magnetic efficiency, or for the cores alone; and sufficient massiveness is required to provide against magnetic over-saturation, loss, and heating, and to furnish a magnetic field of the required strength.

There are usually two pole-pieces, though sometimes a greater number, and the magnets connected with them also vary in number. One magnet may have two or more pole-pieces, or two or more magnet poles be connected with the same pole-piece; the office of the pole-pieces being the proper distribution of the magnetic energy derived from the magnets in the interior space known as the magnetic field, in which the armature rotates, and from which the name, *field magnets*, is derived.

The Series Wound Motor.—In this motor the field-magnet coils are connected in series with the armature and external circuit, as shown in Fig. 3; the entire current employed by the motor traversing each in succession, and generating the magnetism of the field; and the magnets are wound with coarse wire adapted to carry current of the maximum strength required. The arrow within the circle shows the direction of the rotation, the other arrows the direction of the current.

The Shunt Wound Motor.—In this motor there are two distinct internal circuits, in parallel with each other, and connected at the brushes with the external circuit, as shown

PRINCIPLES OF THE ELECTRIC MOTOR. 13

in Fig. 4; the main circuit through the armature coils which carry the principal current, and a shunt circuit through the field magnet coils which are of fine wire and carry 1.5% to

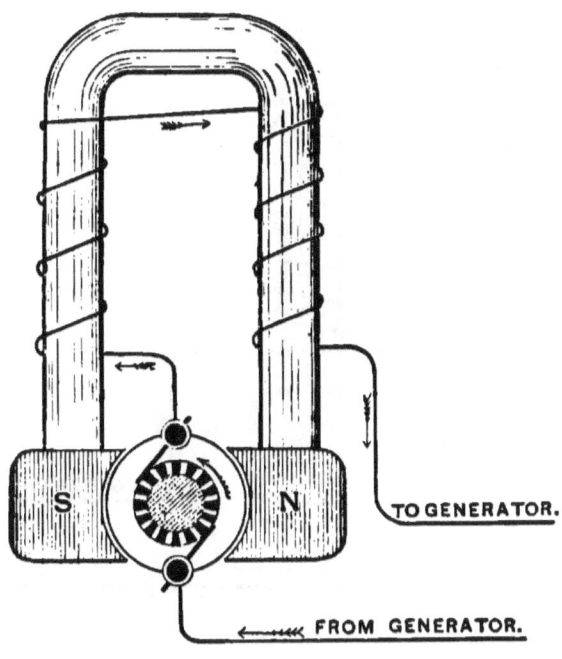

FIG. 3.

20% of the entire current. This shunt current, which is employed exclusively to magnetize the field, leaves the main circuit at the positive brush and returns to it at the negative, as indicated by the arrows, and hence does not pass through the armature as in the shunt wound dynamo. The arrow within the circle indicates the direction of the rotation, as in Fig. 3.

The compound winding is a combination of the series and shunt; coils of low resistance being wound on the field-magnet cores, as well as the high resistance shunt coils, and connected in series with the armature and generator circuits.

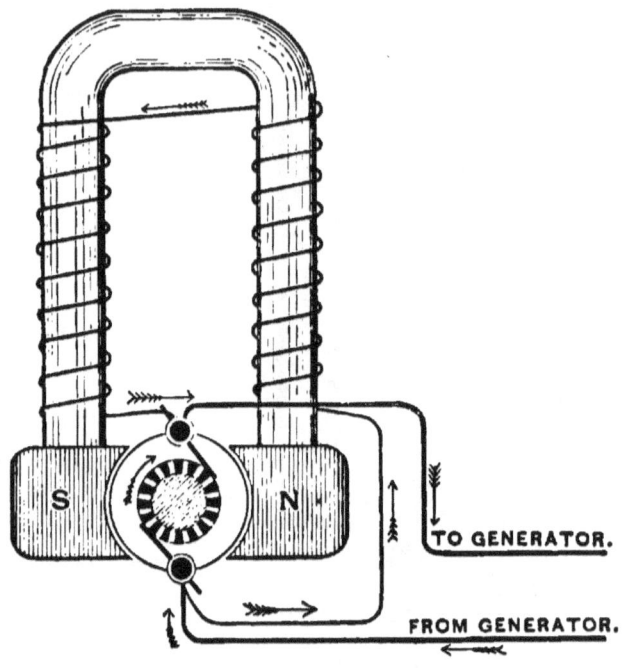

FIG. 4

Operation of the Motor as a Dynamo.—Since the motor may be employed as a dynamo, as already stated, a brief explanation of its operation as such will aid materially in understanding its operation as a motor.

The rotation of the armature, which is the leading feature in the operation of each machine, is produced in the motor in obedience to magnetic polar attraction and repulsion, generated by electric energy; while, in the dynamo, this rotation is produced by mechanical power in opposition to this magnetic influence, and results in the generation of electric energy.

If the motor represented by Fig. 5 were operated as a dynamo by the mechanical rotation of its armature in the direction watch hands move, the winding of its field-magnets being as represented, north polarity would be induced in its right hand pole-piece and south polarity in its left hand pole-piece, the magnetic lines of force traversing the field from right to left. Electric currents would be generated in its armature coils as a result of their rotation in this magnetic field, which, in the exterior wire on the right, would flow from the observer, and in that on the left, toward the observer, and hence in the same direction through a coil wound on opposite sides of a cylindric core, as in the Siemens armature, or in coils wound on opposite sides of a ring core, as in the Gramme; the currents in the latter returning oppositely through the interior wire.

The currents thus generated are of higher potential on the right than on the left, and there is a corresponding difference of magnetic potential generated in the field, and hence they traverse the armature coils, on each side, from the lower to the upper brush, passing into the external circuit by the upper brush and returning to the armature by the lower brush, thus completing the circuit; the main current thus dividing at the lower brush receiving accessions in each branch from the armature coils, and reuniting at the upper brush.

Commutation.—As the position of each coil with reference to this circuit is reversed at each half revolution, as it crosses the neutral line on which the circuit divides, its current is reversed by this transposition, and hence the currents generated in the coils become alternating, while the general course of the armature current, as a part of the main current, is direct. This is effected by the attachment of the coil terminals to the insulated commutator segments, as already described, so that the current entering by a segment in contact with the lower brush, and the currents generated in the coils on each side, must traverse all the intervening coils before gaining an exit by a segment in contact with the upper brush; whereas if the opposite terminals were attached to two undivided metal rings, separately insulated, and each in contact with a separate brush, the currents would pass from brush to brush alternately in opposite directions, and the entire current, external and internal, become alternating. The direct current, produced as above, is composed of the transient currents generated in the armature coils following each other in series, without intermission, through the entire circuit, external and internal.

Polarity and Neutral Line.—The magnetic effect of the currents circulating through the armature coils is to induce

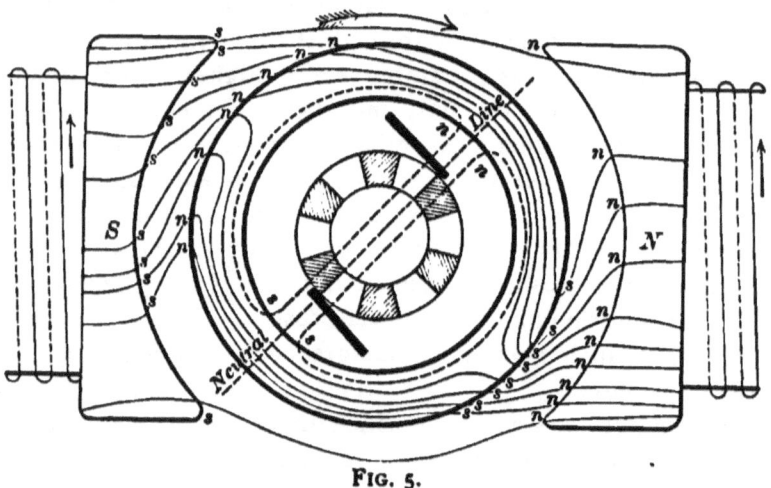

Fig. 5.

in each half of its core, on opposite sides of the neutral line, a north pole on the right and a south pole on the left, in proximity to the corresponding field-magnet poles: and as like poles repel each other, the north field-pole is repelled downward to the lower corner of the right pole-piece, and the south upward to the upper corner of the left pole-piece; the corresponding armature poles being oppositely repelled, the north upward on the right, and the south downward on the left, as in Fig. 5, so that a straight line connecting the field-poles would cross a line connecting the armature poles nearly at right angles ; the latter being the line designated as neutral, since it divides the field magnetically; its position being always diagonal to that of a line joining the centres of the field pole-pieces.

The attraction of opposite armature and field-poles being in the same direction as the repulsion of similar poles tends to produce the same result.

As the currents flow constantly, on each side, in the relative directions described, the relative positions of these armature poles remain stationary, the armature rotating through them, and the polarity of its opposite sections being alternately reversed as they cross the neutral line at each half revolution.

There are also weaker poles induced in the armature core by the field-magnet poles, a south pole on the right, opposite the north field-pole, and a north pole on the left, opposite the south field-pole: but these poles are of minor importance and their influence may be neglected.

Coreless Construction.—As an iron core is not essential to the generation of magnetism or magnetic polarity, both of which are found in coreless coils traversed by the electric current, a dynamo or motor may be constructed without cores, either in its field-magnets or armature, but its energy, both electric and magnetic, is weak, and such construction is employed only where low energy is required for some specific purpose.

Reversal of Rotation.—Reversal of the armature's rotation reverses the current and resulting polarity in both the field-magnets and armature.

Position of the Brushes.—When the brushes make contact on the neutral line all the current generated in every coil passes into the external circuit before the position of the coil is reversed: but if they made contact on a line at right angles to the neutral, the current generated in opposite halves of every coil, during its quarter rotation from the contact line to the neutral, would be neutralized by the reversed current generated during its quarter rotation from the neutral to the contact line, and no current traverse the external circuit. If the contacts were made at intermediate points between these two lines, partial neutralization would result, increasing as the brushes receded from the neutral line, and decreasing as they approached it. By shifting the brush contacts on the commutator in this manner, which is technically termed, giving the brushes *lead*, the generation of electric energy may be regulated and controlled within certain limits.

Operation of the Direct Current Motor.—We are now prepared to consider the operation of this machine as a motor. Since action and reaction are equal and opposite in direction, and since it has been shown that the rotation of the armature in a machine operated as a dynamo is produced by mechanical force in opposition to the reaction of magnetic influence, this reaction must be equal to the mechanical force exerted to overcome it; hence if electric energy derived from an external source, be transmitted in the same direction through a machine similarly constructed, in which this opposing mechanical force is absent, the same relative conditions of magnetic influence will be established, producing rotation of the armature in the opposite direction with a force proportionate to this electric energy; the polarity of opposite armature sections being alternately reversed, as in the dynamo, by reversal of position with reference to direction of current, and the rotation thus sustained.

Effect of Current Reversal on Rotation.—Reversal of current in either the field or armature alone, reverses the relative polarity of these parts, and therefore reverses the rotation; but if the current be simultaneously reversed in each, the relative polarity remains unchanged, and the rotation is not reversed. Hence to adapt the motor to work requiring reversal of rotation, the circuit must be arranged for independent reversal of current in one of these parts, the armature circuit being usually preferred for this purpose.

Polar Rotation.—Since any small section of the armature has opposite poles which remain unchanged till reversed on the neutral line, each half of the armature, on opposite sides of this line, may be considered as composed of a series of such sections having all their north poles turned in one direction and all their south poles in the opposite direction; each small section rotating in this manner till it crosses the neutral line, when it rotates with reversed polarity on the opposite side. Thus, while the main poles, in each half of the armature, occupy fixed positions at opposite ends of the neutral line, the two north poles at one end and the two south poles at the other, the poles of each small section rotate continuously, and are reversed twice during each revolution.

The same conditions of polarity and polar rotation are found in the dynamo, the only difference being, that, in the dynamo, the rotation is produced by mechanical force, and in the motor by magnetic force.

Counter Electromotive Force.—The rotation of the armature in the magnetic field, by whatever means produced, results in the generation of electromotive force; hence the rotation of the motor's armature by magnetic influence, as described, produces this result in the same manner as if it were rotated by mechanical power. But since energy cannot multiply itself, it is evident that the total energy cannot exceed the original energy derived from the source of power; hence it is found, that the energy thus produced is expended in opposing the energy which produces it, neutralizing it in proportion to its relative strength, so that the total amount of energy remains the same, less the loss by friction and inertia.

If a motor be operated by a current from a dynamo, as is usually done, this counter E. M. F. from the motor will oppose that of the dynamo; and as the E. M. F. generated by each machine is in proportion to the lines of force cut by its armature per unit of time, it will vary in proportion to the speed of each armature, in machines of equal size and similar construction. Hence if equality of speed in the two armatures were possible, it would produce equality of E. M. F. in the two machines, and both would stop running; but as the dynamo's current must traverse the motor, and overcome its friction, inertia, and the resistance of its internal circuit, its E. M. F. must always exceed the counter E. M. F. of the motor in the proportion required for this work, and the speed of the dynamo's armature exceed that of the motor's in the same proportion, to supply this excess of E. M. F.

If under these conditions "load" be put upon the motor, that is, if it be applied to the operation of machinery, its speed will decrease in proportion to the load, and its counter E. M. F. decreasing as the speed, will cease, proportionally, to oppose the E. M. F. of the dynamo. But if the dynamo's speed be increased in the same proportion as the motor's load, the motor's speed will remain constant. This increase of speed in the dynamo requires increase of applied power, which is thus developed as load, or work, by the motor; the two machines becoming a combined apparatus for the transformation of power to any extent within the range of their ability. If the motor's load be decreased, its speed and counter E. M. F. will be proportionally increased, and the speed of the dynamo reduced in the same ratio, requiring corresponding reduction of power; the power required thus varying as the load and being constantly proportional to it.

It has been shown that when dynamo and motor are of equal size and similar construction, the motor's speed can never quite equal that of the dynamo; but when the motor is the smaller machine, its speed may greatly exceed that of the dynamo, since its armature must cut an approximately equal number of lines of magnetic force per unit of time, to generate approximately equal counter E. M. F., when running without load; hence its relative speed, under these conditions, must vary approximately in the inverse ratio of its relative size. From which it follows, that the greater the relative size of the motor, especially of its armature, the slower will be its relative maximum speed; and, on this principle, motors may be designed to run at any required speed within practical limits, as more fully explained in Chapter V.

Position of the Poles and Neutral Line.—The rotation of the motor's armature poles, as described, would bring them into close proximity with the opposite field-magnet poles, if it were not for the counter polarity induced by the motor's counter electric influence, by which the main poles are brought to fixed positions dependent on the relative strength of this polarity. The position of the resulting neutral line is approximately the same as in the dynamo when the motor's rotation is in the same direction, but diagonally reversed when the rotation is opposite; and the position of the brushes, in each case, is dependent on the position of this line.

Eddy Currents.—The rotation of the armature in a machine operated either as a dynamo or a motor generates currents not only in its coils but also in its core; and since the core has no connection with the external circuit, these currents circulate round it as eddies without useful effect, consuming energy and generating heat. In laminated cores they are suppressed by the insulated laminæ, which are at right angles to the coils, and their wasteful and injurious effects reduced to the minimum.

In the dynamo these currents and those flowing through the armature coils, being both generated by the armature's rotation, flow in the same direction, and hence the inductive effect of the coil currents tends to neutralize the eddy currents; but in the motor the eddy currents are generated in the same manner as the armature's counter current, and hence flow in the opposite direction to the coil current which produces the rotation, whose inductive effect therefore tends to strengthen them. Hence lamination and insulation are of proportionately greater importance in the motor's armature core than in that of the dynamo.

Loss of Power.—The electric conversion, transmission and application of mechanical energy, as described, involves a certain percentage of loss due to the friction and inertia of the two electric machines, to their electric resistance and self induction, to the resistance of the connecting circuit, and to the dissipation of electric and magnetic energy by imperfect insulation and otherwise, including that due to the formation of the eddy currents. The loss in well constructed machines does not usually exceed 15% of the energy derived from the steam engine or other source of power, and may be considerably less, while that in the connecting circuit varies according to the resistance and insulation; resistance depending chiefly on the distance to which the power is transmitted, and insulation on the attendant external conditions.

Series and Shunt Motors Compared.—Each of these motors has its special adaptation to different classes of work. Since in the series wound motor the entire current is employed to energize both the field and armature, and varies as the resistance through which it passes, as already described, it may be advantageously employed for work requiring great variations of load, and manual regulation to supply the energy, as on street cars, where the maximum energy may be required to start or stop a car, or propel it up an ascending grade while the minimum is sufficient to maintain speed on the level. The current may be adapted to these varying conditions by admitting it through a rheostat, or artificial resistance, controlled by an attendant, by which its strength may be regulated as required.

But where the variations of load are not so great, as in the operation of lathes and other machinery in a shop, and the employment of an attendant would entail too great expense, the shunt wound motor, by its automatic internal regulation, meets these requirements.

In this motor the shunt current, employed to energize the field, as described in connection with Fig. 4, varies in strength automatically as the load on the armature. For this load resists the rotation of the armature, and hence resists and proportionally reduces the strength of the armature current by which the rotation is produced. But as increase of this resistance increases the electromotive force or potential difference between the opposite terminals of the divided circuit, at the brushes, it increases the current strength in the shunt, since the resistance in this branch remains constant, and hence increases the magnetic energy of the field and proportionally restores the strength of the armature current; thus maintaining approximate constancy of strength in the armature current with variation of load, since decrease of resistance reverses these results.

As the shunt circuit is in parallel with the armature circuit, the polarity of the field with respect to that of the armature is the reverse of its polarity when the two circuits are in series, and hence the rotation is the reverse; and since it has been shown that the rotation of a series motor is opposite in direction to that of a connected dynamo when the current traverses the armature in the same direction in each machine, the rotation of a shunt motor, with the same direction of armature current, must be in the same direction as that of the dynamo.

Constant Current and Constant Potential Circuits and Motors.—An electric circuit may be constructed to carry a current of constant strength with a varying electromotive force, or potential, or a current of varying strength with a constant potential; and the motors employed on each must be adapted to the requirements of the circuit.

If a circuit is constructed to carry a current by which motors of different sizes, arranged in series, are operated, and arc lamps interspersed among them, at different points lighted, as illustrated by Fig. 6, in which the squares represent motors and the circles lamps, this current must be

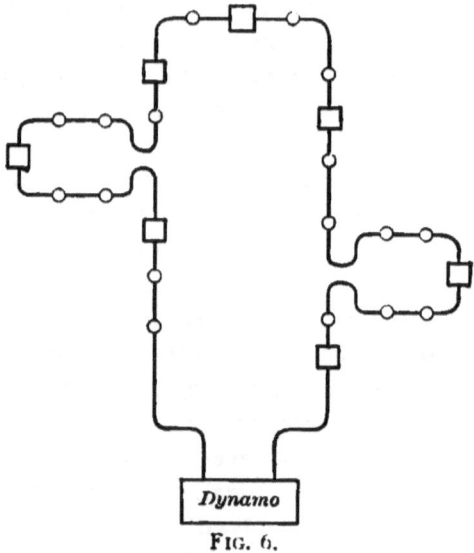

Fig. 6.

maintained at a constant strength throughout the entire circuit of motors and lamps.

As the number of motors and lamps in use simultaneously is liable to continual variation, the resistance encountered by the current varies in like proportion; hence, to maintain constancy of current strength, the potential of the dynamo, or combined dynamos, at the central station, by which the current is impelled, must be correspondingly varied.

The motors employed on such a circuit must all be series wound and each employ a given number of watts of energy proportioned to its capacity and varying load : a one-horse motor, for instance, requiring 746 watts, at maximum load, and a two-horse, 1492 watts, this quantity being reduced in each in proportion to reduction of load. But as the current flowing through each motor has the same strength, which, for instance, may be ten amperes, the two-horse motor must be constructed to produce a maximum potential, or E. M. F., double that of the one-horse, to produce double the energy, and the potential in each must be varied automatically or manually in proportion to variation of load and required energy.

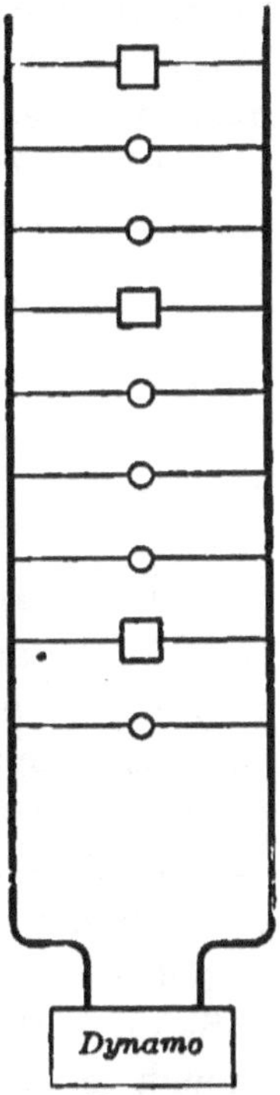

But if the circuit is constructed to carry a current which flows through two leading mains connected by branch circuits, on some of which are placed motors and on others incandescent lamps, as shown in Fig. 7, the current strength in the mains will vary as the number and size of the motors and lamps in use simultaneously, while the potential by which the current is impelled remains constant.

For the resistance of the mains and connected dynamo being constant, the strength of current flowing through the mains with a constant potential must vary as the resistance encountered by the current in the branch circuits ; and this resistance varies inversely as the number of branch

FIG. 7.

circuits in use simultaneously and the strength of current in each. Suppose that no lamps are lighted and only a single motor in operation, that the potential or E. M. F. is 100 volts, and that this motor requires a current of 10 amperes; the resistance of the branch circuit, including that of the motor and its counter E. M. F., must be made ten ohms, for dividing the numerical value of the volts by that of the ohms, we have $\frac{100}{10} = 10$ amperes. If now a second motor of the same size be put in operation under the same conditions, the required current strength for each being the same, the resistance in each branch must be made the same. But as the current between the mains has now two parallel conductors of equal capacity, the total branch resistance is reduced one half, and hence the total current strength, with the same potential, doubled; for, dividing as before, $\frac{100}{5} = 20$ amperes, and this current being divided equally between the two motors, gives each ten amperes. In like manner the current strength for any larger number of motors of different sizes is made proportional to the requirements of each.

The constant resistance of the mains and dynamo is too small to be practically appreciable on such a circuit when only a small number of motors and lamps are in use simultaneously, but when the number is large, and the total branch resistance becomes correspondingly reduced the ratio of the constant resistance to the branch resistance is proportionally increased, requiring corresponding increase of potential at the dynamo to supply the required current.

Shunt motors are generally preferred on such circuits, though series motors may also be employed; and in both cases the current is admitted to the motor through a rheostat, by which its strength is regulated as required.

The application of the terms *arc light* and *incandescent* to

PRINCIPLES OF THE ELECTRIC MOTOR. 27

motors employed on circuits so designated, which has become common, is misleading and manifestly improper.

The Rheostat.—The rheostat is an apparatus for regulating current strength by varying the resistance through which the current must pass when admitted to a motor or other electric apparatus. Fig. 8 shows an instrument of this

FIG. 8.

kind constructed by the Detroit Motor Co. which may be selected as a representative of such instruments when designed for motors.

A wooden case incloses a number of German-silver resistance coils connected together in series. Attached to its insulating cover are nine brass stops slightly separated, and arranged in an arc, as shown, each, except No. o, connected with the resistance coils. A jointed switch, pivoted near its centre, can be rotated by an insulating handle so that its upper end can make contact with each stop, while its lower end is in contact with segment K. At the right of K is an insulating stop, C, and just below it a brass stop, B. Binding posts A and E are for connection with the dynamo, or other source of electric energy, and D, H and F, for connection with the motor.

When the connections are made and the switch is in the position shown, its upper end resting on the blank stop O, and its lower end on the insulating stop C, an insulating strip, connected with it, separates the stop B from a copper spring connecting B with A, and no current can traverse the rheostat; but when the switch is moved into connection with stop 1 above, and segment K below, the spring closes the connection between A and B, and current begins to flow to the motor.

The current entering by post E, passes by an inside connection to segment K, thence through the switch and resistance coils to stop 8, and thence by an inside connection to F. Here it divides, the armature current going direct to the motor, which is shunt wound, and having traversed the armature coils, returns to D, thence by an inside connection to B, thence through the connecting spring to A, and thence back to the dynamo. The field current passes from F by an inside connection to an electromagnet, one of whose poles is shown at L, the other being under the switch pivot, and having traversed its coils, passes by an inside connection to H, and thence to the field-coils of the motor, which, having traversed, it reunites with the armature current and returns to D, and thence back to the dynamo by B and A.

The relative direction of the current is of no importance. It may enter the rheostat by A if preferred, and traversing the circuit in reverse order, return to the dynamo from E; the result being the same with either direction.

When the switch is on stop 1 the current must traverse all the rheostat coils from 1 to 8, encountering the full resistance and hence having very limited strength; but as the switch is moved toward 8, from stop to stop, the current increases in strength with the decrease of resistance, and the speed of the motor increases in like proportion, till the switch rests on stop 8, when, the entire resistance being cut out, the current attains its maximum strength and the motor its maximum speed.

When the switch is in this position, an iron armature, attached to its under side, makes contact with the poles of the electro-magnet referred to, and holds it on stop 8 in opposition to a spiral spring at the pivot which tends to force it back toward stop 0. But when the motor is to be stopped, the lower part of the switch, which is attached to the upper by a knife hinge at the pivot, is moved to the right without disturbing the upper part, and the insulating strip being pushed under the spring at B, the electric circuit is broken, the magnet demagnetized, and the upper part of the switch thrown back by the spring to its first position. Thus when the circuit is broken by accident or otherwise, the switch always flies back automatically to stop 0, and can never rest on any intermediate stop between 0 and 8, or be left there by the negligence of the operator. The importance of this becomes evident when it is considered that the resistance coils in this rheostat are of limited capacity, and intended to be employed only in starting the motor, and are therefore in use only while it is acquiring its speed, after which the current is regulated by the relative proportions of load and counter electromotive force, as already described. Hence if the switch should be allowed to remain too long on any intermediate stop, the result would be the heating and destruction of the resistance coils through which the current passed.

In stopping the motor, if the switch were not jointed, it would have to be moved from 8 to 0 before the circuit could be broken, which would produce a spark, by the interruption of the current, at each space between stops, resulting in heating, oxidation, and permanent injury to the contacts. But the joint permits the circuit to be broken below, as described, before the upper part of the switch is released from stop 8, and thus prevents the sparking.

The construction of rheostats varies with the specific use for which each is intended, and embraces many excellent designs; the essential parts in each being the coils, switch, and stops. Some are intended to carry much stronger currents than others, and require coils of corresponding capacity. Others are intended to be under the constant control of an operator, and do not require the automatic movement of the switch described above.

Rheostat Connections.—The terminal connections with the rheostat and mains are shown in Fig. 9. The two mains above are on a constant potential circuit connected with one or more dynamos, and may issue from a central station and carry sufficient current to supply a large number of motors and incandescent lamps arranged on derived circuits in parallel, as in this case.

Each of the vertical wires connects with a separate main, so that the current traversing the motors flows across from one main to the other; and hence any interruption on this branch circuit, either by the stoppage of the motor, or the opening of a switch in the branch cut-out above, does not interfere with the supply of current to other motors and lamps.

The safety fuses are short pieces of soft metal alloy, which are placed in the circuit, as shown, and fuse at a

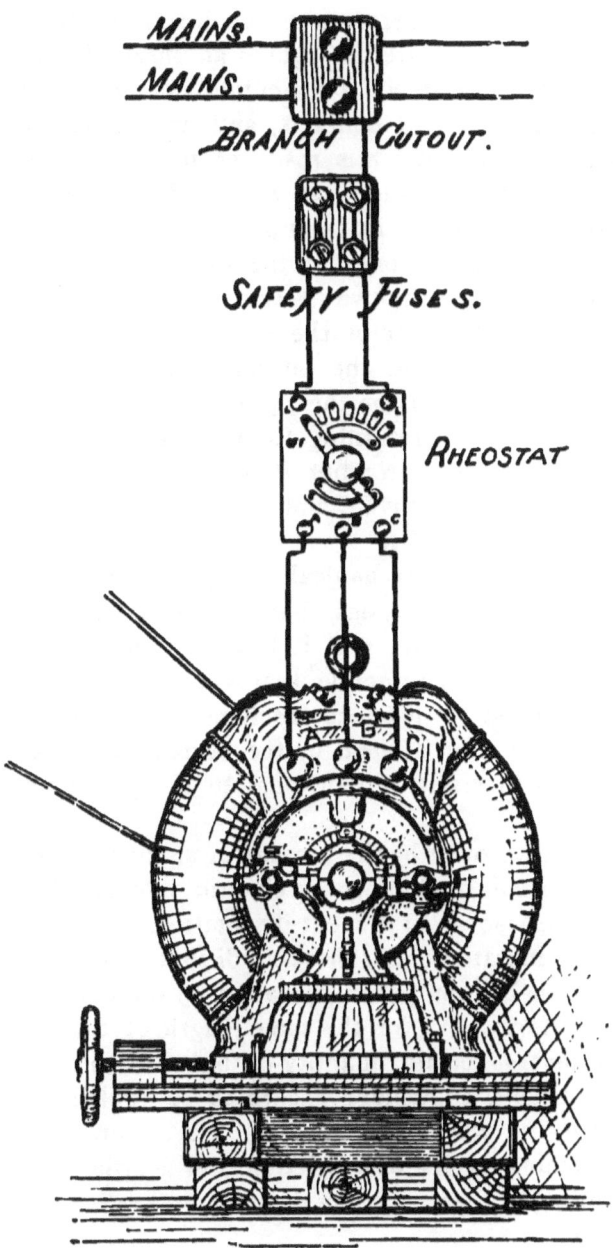

Fig. 9.

comparatively low temperature and open it, when, from any accident, the temperature approaches an unsafe degree.

The rheostat here shown has a curved contact block just below the upper row of contacts, and another just above the lower one; the switch is made of insulating material and carries a contact block at each end; the upper one connecting each contact in the upper row with the curved block below it, as it moves to the right or left, and the lower one connecting the two lower contacts. The current enters the rheostat by one of the upper terminals, LL, and leaves by the other, and the interior connections are so arranged that if it enters on the right it goes to both armature and field by AA, returning from the armature by CC and from the field by BB; but, if it enters on the left, it goes to the armature by CC and to the field by BB, and returns from each by AA.

Electric Heat and Mechanical Energy.—It is a well established principle of physics, that mechanical energy may be converted into heat; and it has been ascertained, in accordance with this principle, that when the electric current is employed to produce mechanical energy by the motor, the generation of heat by the current, in traversing the circuit, is inversely proportional to the production of this energy.

This may be proved by the passage of an electric current through a circuit including an electric motor, with the motor clamped so as to prevent its rotation. If the quantity of heat generated under these conditions be ascertained, and the motor be then put in operation by the current, and made to perform work for the same length of time, it will be found that a certain quantity of heat has disappeared from the circuit, equivalent, according to Joule's law, to the quantity of work performed by the motor, including the overcoming of its own friction and inertia; Joule's equivalent being that 42,000,000 ergs in work are required, when converted into heat, to raise the temperature of 1 gramme of water 1° C. It thus appears that electric heat may be transformed into mechanical energy.

If now the experiment be reversed, by first putting the motor in operation by the current for a given time, and then suddenly stopping it by a brake, and allowing the current to traverse it for the same length of time, it will be found that the heat developed is equivalent to the mechanical energy which has disappeared; proving that mechanical energy may be transformed into electric heat.

This experiment could be performed with a small motor, inclosed in a water-proof envelop and immersed in a vessel of cold water, in which the electric heat was properly confined, and indicated by a thermometer; the heat generated by a weak current for a limited time, during the stoppage of the motor, being absorbed by the water, thus preventing injury to the motor.

The application of this principle in the operation and construction of motors is of the highest importance. If, under ordinary conditions, a motor, while in operation at full speed, should be suddenly stopped by a brake or otherwise, while the current were still permitted to traverse its coils, the energy which disappeared as work, during the stoppage, would appear as heat, and the result would probably be the charring and consequent destruction of the insulating material of the armature, and possibly that of the field-coils also.

Likewise, if, in putting a motor in operation, the full current should be turned on before the armature had acquired corresponding speed, heat would be generated in a similar manner, with similar results; hence the importance of admitting the current slowly, through a rheostat, in the manner already described.

It is evident from this that anything in the construction or material of an electric motor, which generates unnecessary heat, proportionally reduces the mechanical energy. Hence, in designing a motor, this principle should be strictly observed, in the choice of material, and the relative proportions, arrangement, and construction of the different parts.

Motor Designing.—Since a motor may be regarded as a reversed dynamo, the same principles of construction apply to both; hence the motor should, like the dynamo, be constructed with special reference to the generation of electromotive force when similarly operated; and the machine which accomplishes this most efficiently is the best either as a dynamo or motor. The best motor is therefore the highest generator of counter electromotive force.

The kind of winding to be adopted, whether shunt or series, must be determined by the use to which the motor is to be applied, and the kind of a circuit, whether constant potential or constant current, on which it is to be employed; the shunt wound motor being adapted to the former, and the series wound to the latter, as already shown.

If it be required to design a series motor of a given horse-power, we may properly begin with the armature, on which the relative construction of the other parts largely depends. The choice of armatures lies chiefly between the Gramme and the Siemens, each of which has its special advantages and defects. The Siemens is generally preferred to the Gramme on account of its solid mechanical structure, higher magnetic efficiency, and smaller proportion of idle wire. Its principal defects are the accumulation of idle end wire, with numerous crossings liable to short circuits unless carefully insulated, and also to heating; the limitation of its diameter, which cannot be increased to the same extent as the Gramme without excessive increase in the relative proportion of end wire; and the greater difficulty of securing the wire against displacement by the rotary force on a relatively long cylinder, than on a ring which can be made relatively short, like the Gramme.

The Gramme armature can be increased in diameter to any desired extent without increase in the relative proportion of end wire, giving slower speed in proportion to electromagnetic efficiency, and a shorter ring, on which the coils can be more securely wound, especially if confined by core projections. Its open structure gives better ventilation and heat radiation than the Siemens, in which tubular openings through the core are often employed for this purpose. Its chief defects are interior wire, which, according to the best evidence, is comparatively idle, like the end wire, serving only as a conductor, and taking no part in the generation of electromotive force, and hence increasing the weight, cost, and electric resistance without corresponding increase of electric efficiency; the relatively greater thickness of mass and crowding together of this interior wire as compared with the exterior, due to the reduced interior circumference, and space occupied by the ring supports; less solidity and greater cost of core construction; and less magnetic efficiency, the two halves of the core being traversed by the magnetic circuit in parallel, while the Siemens core is traversed by a single circuit.

The generation of electric energy varies in proportion to the total magnetic force developed by the joint action of the armature and field; and this is determined by the relative quantity of iron in the cores of each, its kind and quality, its magnetic saturation by the coils, and the number of times the lines of magnetic force are cut by the armature per unit of time, which corresponds in number to the reversals of current, and is dependent on the armature's speed and relative diameter, and the number of field-poles.

The armature core should be composed of thin plates of the best soft, wrought iron, insulated from each other with tissue paper; and, if of the Siemens type, ventilated by tubular openings parallel to the shaft; connected, at short intervals, with the surface by air spaces. Its diameter and length depend largely on the horse-power required, as well as on the speed and type of armature. If the Siemens type be chosen, in which there is only a single magnetic circuit, the entire mass in longitudinal section, deducting the ventilating space, can be made the basis of calculation; but if the Gramme type is preferred, in which there are two magnetic circuits in parallel on opposite sides of the ring, the calculation is limited to half the mass in longitudinal section; that is, to a radial section of one side of the ring parallel to the shaft.

When a core of either construction is wound with wire of suitable size, and in sufficient quantity to produce magnetic saturation, it is found that the E. M. F. generated by each convolution, with a rate of speed producing 1000 current reversals per minute, is equal to $\frac{3}{100}$ of a volt per sectional square inch of core; estimating by the entire section of the Siemens or the half section of the Gramme, but including only half the number of convolutions in each, since the coils on both sides generate E. M. F. in parallel, in each armature.

The generation of E. M. F. is the same in the exterior and interior convolutions; the magnetic induction in each varying directly as the length of wire in the convolution and inversely as the square of its distance from the core, and each of these factors varying directly as the other; hence the reduced induction due to distance from the core in the exterior convolutions is compensated by greater length of wire, while the reduced induction, due to reduction of wire length, in the interior coils, is compensated by nearness to the core.

The number of current reversals per minute depends on the speed of rotation and number of poles in the field. In a two-pole field, the lines of magnetic force being cut twice at each revolution of the armature, there are two current reversals; hence in a slow speed, two-pole armature having, say, 1000 revolutions per minute, the number of current reversals per minute would be 2000, and the generation of E. M. F. per sectional square inch of saturated armature core $\frac{8}{100}$ of a volt; while, in a multipolar field, the relatively greater number of poles would produce corresponding increase in the number of reversals and generation of E. M. F. From which it is evident that reduction of speed, with a given size of armature, can be obtained without reduction of E. M. F. by increasing the number of field-poles; minimum speed with maximum efficiency being always desirable. But the construction of multipolar, direct current motors is more complicated than that of bipolar motors, and the distribution of magnetic force in the field in equal proportion between the several pairs of poles, difficult to adjust properly.

The size of wire required for the armature coils depends on the current to be carried, which being equally divided between the two armature circuits in parallel, gives half the current carried by the mains to each; equal, on a ten-ampere circuit to five amperes. The carrying capacity of wire is reckoned by circular mils in cross section; a mil being $\frac{1}{1000}$ of an inch in length, and a circular mil, the area of a circle whose diameter is one mil. For pure copper wire, in coils properly ventilated, 800 circular mils per ampere is considered a fair carrying capacity; which, for 5 amperes, gives $5 \times 800 = 4000$ circular mils in cross-section, which, according to the wire table on page 6, gives No. 14 B. & S. gauge as the nearest corresponding size; this, as shown by the table, is 64 mils in diameter, equal to $\frac{64}{1000}$ of an inch, for naked wire, or about $\frac{78}{1000}$ of an inch, including insulating material. Hence since 1 inch $\div \frac{78}{1000}$ inch $= 13\frac{15}{100}$, the winding gives about 13 strands to the inch.

The quantity of wire which can be wound on an armature core is limited by the space intervening between the core and field pole-pieces; this should be as narrow as possible, not exceeding $\frac{1}{2}$ an inch, since the magnetic resistance of the air, copper wire, and insulating material by which it is occupied is very great and generally regarded as equal to that of air alone. Allowing $\frac{1}{16}$ of an inch for clearance between the coils and pole-pieces, and $\frac{1}{16}$ for insulation of the core, which should be well covered with mica and oiled paper before winding, the thickness of the coils is limited to about $\frac{3}{8}$ of an inch, or 5 layers of No. 14 wire.

If the armature core is constructed with projecting teeth, between which the coils are wound, and which can therefore come within $\frac{1}{16}$ of an inch of the pole-pieces, it is claimed that the magnetic resistance is reduced, though the average distance between armature core and pole-piece remains the same, since corresponding increase of depth is required in the depressions.

The length of wire in the coils is determined by the size, type, and speed of the armature, and number of current reversals per minute; calculated with reference to the required horse-power of the motor by multiplying the E. M. F. in volts into the current in amperes, which gives the electric energy in watts, which being divided by 746, the number of watts per horse-power, gives the result in any given case; and the weight and resistance of the wire can be ascertained from its length and gauge by the table. The quantity of wire should not exceed that required to saturate the core, as any excess increases the resistance, weight, and expense, without corresponding useful effect. Perfect saturation is not strictly necessary, though desirable as a matter of economy in weight and expense of core; but the quantity of iron should always be sufficient for the maximum saturation. It is highly important that the winding should be perfectly even and regular, like thread on a spool, to produce proper electromagnetic action; and that the armature should be accurately balanced to insure smooth, even rotation and freedom from noise.

The field-magnet cores may be made either of cast-iron or wrought. The magnetic permeability of wrought iron being about 80 per cent greater than that of cast, it is to be preferred, unless its comparative cost exceeds the comparative advantage of quality; but as its use for the yoke and pole-pieces is of less importance than for the cores, they are generally made of cast-iron. The cross-section of each field-core should be $\frac{1}{4}$ to $\frac{1}{3}$ greater than that of the armature core, if made of wrought-iron, like the latter, or not less than double that of the armature core, if made of cast-iron. The cross-section of the yoke should be at least as large, and there is no objection to making it larger, as is often convenient, when the base is utilized for the yoke.

The pole-pieces should have larger cross-section than the field cores, and each, in a bipolar motor, should cover 25 to 30 per cent of the armature surface, so as to reduce the magnetic resistance between field and armature to the minimum. But the air space between their edges should be sufficient to resist magnetic cross-leakage outside of the armature; and as this space embraces the neutral zone, in which the brushes make contact with the commutator, proper adjustment of the brushes with reference to the neutral line is difficult if it is too narrow.

The magnetic circuit through the cores and yoke should be as short and as free from sectional joints as possible, to reduce its resistance to the minimum; and where joints are required, the fitting should be strictly accurate; and all projecting angles, liable to magnetic leakage, should be avoided. Hence one of the best forms for cores and yoke is that of a cylinder of uniform size throughout, bent into the shape of a horse-shoe, or U, and terminating in the pole-pieces.

As the two sets of field-coils are usually connected in series, they carry the entire current, which, in the case assumed, is 10 amperes; hence they require much larger wire than the armature coils. Multiplying the 10 amperes by 800 circular mils, as before, we obtain 8000 circular mils as the required cross-section; to which No. 11 wire is the nearest corresponding size given in the table, but No. 10 is considered a safer size, to insure against heating. It is estimated that with $\frac{1}{2}$ an inch space between armature core and pole-pieces, 1000 ampere feet per square inch of field-core in cross-section is required to obtain magnetic saturation; that is 1000 feet having a carrying capacity of 1 ampere, or a less number having a proportionally greater carrying capacity. Hence, multiplying the square inches in cross-section by 1000 and dividing the product by 10 amperes, gives the required number of feet of wire, from which the weight and electric resistance can be obtained by the table.

In designing a motor for any given efficiency, allowance must be made for the fall of potential in the field and armature coils respectively, and the E. M. F. which it is designed to develop be made correspondingly greater. This fall of potential can be ascertained from the resistance of the wire, as given in the table, estimating by half the resistance of the armature wire, in which there are two circuits in parallel, and the full resistance of the field wire. Allowance must also be made for the difference between the theoretical efficiency obtained by calculation, and the actual practical efficiency developed, which may be estimated at about 25 to 30 per cent less.

The relative polar strength between field and armature is of great importance, and must not be overlooked. Proper construction, according to the principles here given, should produce the requisite magnetic balance; but if the polarity of the armature is relatively too strong, its poles will be abnormally deflected by their magnetic reaction on the field-poles, increasing the contortion of the magnetic circuit through its core, between the field-poles, and hence lengthening it and thus increasing the magnetic resistance, and deflecting the neutral line from its normal position.

In constant potential motors the relative resistance of the shunt wound field-coils should be at least 324 times that of the armature coils to insure maximum efficiency; and the quantity of iron in both armature and field cores should be relatively greater than in constant current motors, and the degree of magnetic saturation proportionally less, especially in the armature, to prevent its abnormal reaction on the field-magnets.

But little need be said in regard to the commutator, which has already been fully described. Its segments, or bars, should be made of the best hard copper, and the surface turned perfectly smooth, to reduce the friction and wear of the brushes to the minimum. The bars should be securely confined with steel supporting collars mounted on the shaft, and the insulation, between the bars and from the collars, should be ample.

As alternating current motors are still in the experimental stage of development, the method of designing them will be better understood from the examples given in the next chapter than from any general rules which could be given.

Only the simple, elementary principles of motor designing are here given, from which the reader can get some general idea of how it may be done; a more full exposition of which, including its mathematical formulæ, may be found in works on electrical engineering.

CHAPTER III.

STATIONARY MOTORS.

The practical application of the principles set forth in the preceding chapter to the construction of motors can be best understood by giving examples of such construction, and for this purpose a few of the motors which have come into general use, and whose efficiency has thus been tested have been selected.

The Excelsior Motor.—The standard motor of the Excelsior Electric Co. is series wound, and is the invention of Wm. Hochhausen. Its construction will be understood from Fig. 10. Its two field-magnets are mounted vertically on a cast-iron frame whose base forms the connecting yoke to which the cores are bolted. These are also of cast-iron, each forming a single casting with its pole-piece.

The armature, mounted on projecting supports between the pole-pieces above, is of the Siemens construction, and has the laminated wrought-iron core already described. Its coils are of No. 16, insulated, copper wire, and their terminals are connected to the commutator segments with iron wire, which has greater tenacity and firmness for this purpose than the copper wire of the coils.

The commutator is insulated from the shaft by an air space, and supported on brass collars at each end, from which it is insulated; and its segments are insulated from each other by strips of mica, forming a compact, laminated structure.

The field-magnets are wound with No. 14 insulated copper wire, and the coil terminals brought up outside each pole-piece, as shown, and connected to small, copper plates, contained in two boxes *A* mounted on a switch-board above the armature, one over each pole-piece. These plates are placed vertically, in a compact manner, in each box, which is open on the inside, leaving their interior ends exposed,

FIG. 10.

and are insulated from each other, and each set insulated from the opposite set by the wooden switch-board. A pair of brushes known as the field contact brushes, traverse the space between the boxes and make contact with the ends of the plates. They are constructed with two copper-tipped brass springs, joined together at *B* on the right and connected to a brass sliding-bar underneath; the copper tips on the left pressing against the plates. The sliding-bar is attached to a rod *C* which extends to the right and is hinged to a vertical lever *D*, which is also hinged below and connected a short distance above the hinge, by a ball and socket joint, to a sleeve mounted on the armature shaft and projecting from a box *E* containing a governor, by whose action it is forced to the right, pulling the connected rod against the tension of a brass spiral spring *F*, shown above the commutator.

The governor is constructed with two L-shaped brass castings, each hinged at its angle to a collar mounted on the armature shaft; the short spur of the L carrying a pin which fits into a groove in the sleeve referred to, while the long part, which is a flat, heavy plate, extends along the sleeve to the right. These castings are attached on opposite sides of the shaft, and their right ends connected together by two spiral, steel springs, one on each side. When the shaft rotates, these ends diverge by centrifugal force in opposition to the tension of the springs, and the sleeve is forced out, producing the movement of the rod, as described.

The current traverses the coils of the two field-magnets in two separate, parallel circuits of equal capacity, the terminals of which form a junction at the right of the two wires *G* and *H* which project above the armature, and are connected with the left wire *H*. If the current enters by this wire, it divides at the junction, and having traversed the field-coils from right to left, returns through the two opposite sets of contact plates to the field contact brushes, reunites at the sliding-bar, and passes through it to the front wire *I* leading to the lower commutator brush, and having traversed the armature coils, returns by the upper commutator brush, and rear wire *J*, and passes out by the right hand wire *G* above.

Since the current traverses the field coils from right to

left, it is evident that when the field contact brushes make contact at the left ends of the boxes, all the coils are in circuit, and the electric energy supplied to the motor is at its maximum ; but when, by the action of the governor, the brushes are moved to the right ends of the boxes, these coils are all short-circuited, and the energy reduced to its minimum ; while, with the brushes at any intermediate point, the number of field-coils in circuit, and the resulting energy, is in proportion to the distance to which the brushes are moved to the right. Hence as the speed and resulting action of the governor varies as the load, the number of coils in circuit and resulting energy is thus made to vary automatically as the load, while the strength of current flowing through the motor remains constant.

By varying the tension of the sliding-bar spring by means of the set screws shown at K, the distance to which a given speed will move the brushes may be correspondingly varied, and the supply of electric energy regulated as required by the work. In like manner the action of the governor may be varied by the use of springs of greater or less tension, if the above regulation is found insufficient.

The motor is started and stopped by the switch whose handle L is shown at the left end of the supporting plate above. This switch, when closed, short-circuits the terminals G and H of the wires by which current enters and leaves, and therefore stops the motor, but, when open, the current must pass through the motor and put it in operation. The field-contact brushes are pulled to the right before starting, so as to short-circuit all the field-coils. The switch is then opened and these brushes returned slowly to the left, so as to admit the current gradually to the field-coils, while the armature is attaining its speed.

If the current enters by the right hand wire G it traverses the circuit described in reverse order and leaves by the left hand wire H, producing rotation in the same direction as before, so long as the relative direction of current through field and armature is the same. To reverse the

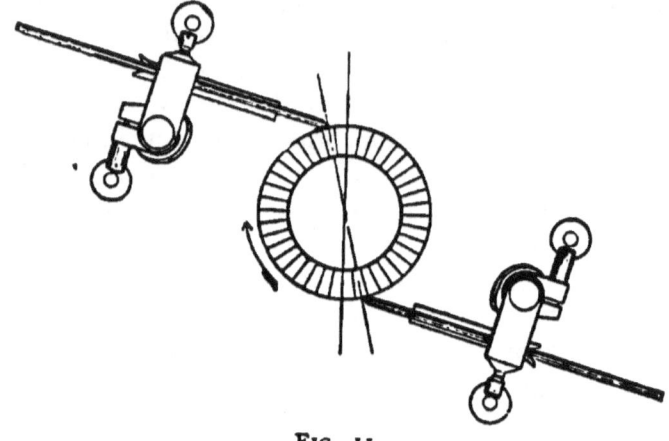

FIG. 11.

rotation, the relative positions of the commutator brushes must be reversed, as shown in Figs 11 and 12, which is done by slipping each off its pin and turning it over before

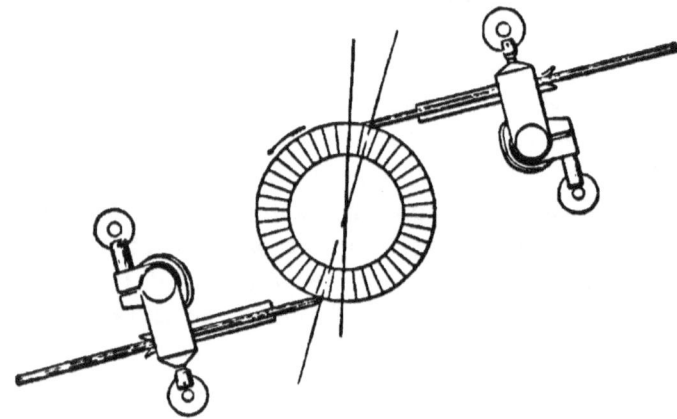

FIG. 12.

replacing it ; the relative direction of the current through the armature being thus reversed, the rotation is reversed.

The brush contacts should be on the neutral line, which makes an angle of about 15° with a vertical line; its relative position being reversed by reversal of rotation, as shown.

The oil cups are placed under the shaft bearings, and

FIG. 13.

the oil supplied to the bearings by strips of felt, through openings in the babbit metal boxes.

The range of these motors in capacity is from $\frac{1}{2}$ to 15 horse-power.

The Edison Standard Motor.—The Edison standard motor is represented by Fig. 13, and is of the same construction as the Edison dynamo. It is shunt wound, has two field-magnets mounted vertically above the armature; their cores are of wrought-iron, connected above by a massive wrought-iron yoke and attached below to cast-iron pole-pieces, bolted to a cast-iron base, but magnetically insulated from it by zinc castings. The armature is supported on cast-iron standards, is of the Siemens type, and has the usual laminated sheet-iron core. The brushes are usually of copper, but carbon brushes are also employed.

Each shaft bearing is provided with two revolving rings, one at each end, fitted loosely to openings in the upper part of the bushing, so as to rest on the shaft and rotate with it, and dipping into an oil reservoir below, from which the oil is brought up by adhesion, and after distribution over the bearing, returns to the reservoir by a central groove and lower opening, to settle and cool; constancy of supply being thus automatically furnished with no other care than renewal.

The switch-board, carrying the circuit terminals and switch for opening and closing the circuit, is shown mounted on the yoke. The current is admitted through a wall rheostat in the usual manner.

These motors range in energy from $\frac{1}{4}$ to 210 horse-power, in weight from 90 to 31,790 pounds, and in armature speed, from 2100 revolutions per minute, for the smallest, to 360 for the largest.

The Edison Small Motor.—The great demand for small motors has led to the construction by Edison of a motor specially designed to meet the requirements of various kinds of small machinery, by combining high efficiency with slow speed and minimum care. It is represented complete by Fig. 14, and its principal parts in detail by Figs. 15, 16 and 17. The construction of the magnetic field is shown in Fig. 15. It consists of a single electromagnet, mounted near the motor's base for steadiness, and supported between

STATIONARY MOTORS. 49

Fig. 14.

two standards which carry four pole-pieces, as shown. The core, standards, and pole-pieces are of wrought-iron, and the pole-pieces, in some of the motors, laminated, as shown, and bolted to the standards, while in others they are solid projections from the tops of the standards.

The armature, shown in Fig. 16, is a Gramme ring with a laminated sheet-iron core having projecting teeth between which the coils are wound, and which hold them firmly in place. It is mounted between the four pole-pieces, two of

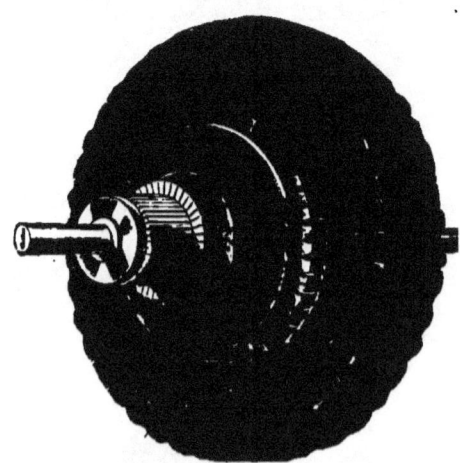

FIG. 16.

which are on each side, parallel with the core, and as close to it as safe clearance will permit. This arrangement reduces the magnetic resistance between field and armature to the minimum, gives a direct course for the passage of the magnetic energy, and suppresses eddy currents and resulting heat in the most perfect manner, especially in the laminated pole-pieces. The coil terminals of the armature are connected to the commutator segments with german-silver wire, the high resistance of which reduces sparking at the brushes, to which small motors are especially liable. The coils being wound separately, an injured coil can easily be removed and replaced without disturbing the others.

As the construction gives the armature maximum diameter in proportion to electric efficiency, its speed is reduced to the minimum; ranging, at full load, from 2000 revolutions per minute, for the $\frac{1}{12}$ horse-power, to 1300 for the $\frac{1}{2}$ horse-power. The brushes are set in a fixed, and approximately radial position, on the neutral line, and permit

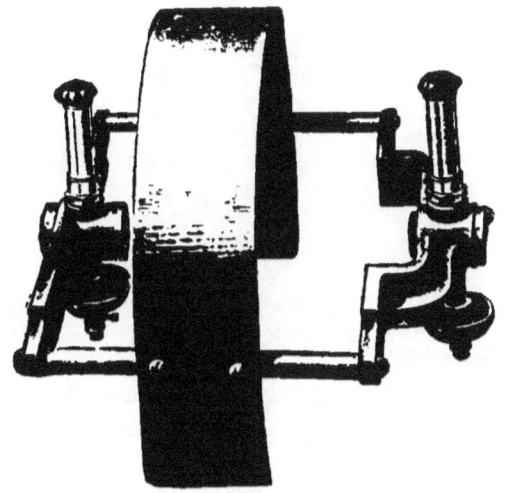

FIG. 17.

reversal of rotation, when required, without change of position.

The brass frame, shown in Fig. 17, is bolted to the standards, and supports the armature, with its guard and lubricating apparatus, as shown in Fig. 14. Brass bushings, easily replaced when worn, are employed for the bearings, and grease for lubricating, which requires less care than oil, and is less liable to spatter and soil; sufficient heat being generated by the friction to melt and supply it, as required, from the holders above the bearings.

A mahogany switch-board, shown in Fig. 14, carries the

starting-switch, circuit connections, and safety fuse, and the motor is mounted on a neat wooden base. The range of energy at full load, with 120 volts E. M. F., is from $\frac{1}{18}$ to $\frac{1}{4}$ horse-power, and the weight from 15 to 38 pounds. The $\frac{1}{2}$ horse motor is shunt wound, and the smaller sizes series wound.

The C. & C. Standard Motor.—The principal motor of the C. & C. Electric Motor Co. is represented by Fig. 18, and

FIG. 18.

is shunt-wound. Its field-magnets are constructed in the form of a circle, the two cores, which are of wrought-iron, being bolted to cast-iron pole-pieces above and below, the lower pole-piece and base being cast in one piece. This construction gives great compactness and reduction of size in proportion to efficiency, bringing the field-magnets into the closest possible proximity to the armature. It produces what are known as consequent poles; the two opposite poles of each magnet meeting at the centre of each pole-piece. The parts are all accurately fitted together, and have few projecting angles liable to magnetic leakage. The pole-pieces, which are of much larger cross-section than the cores, embrace about 280° of the circumference of the armature. The two sets of field-coils are in series with each other, and are protected by an outside covering of canvas and rubber tape, varnished.

The armature is of the Siemens type, with laminated core held together at the ends by iron arbor plates, keyed to the shaft and bolted together. It is mounted on a steel shaft having an enlarged interior diameter, and is turned smooth before winding and thoroughly insulated from the coils with mica and oiled paper. The coils are wound with wire of large size, and proportioned to carry an excess of current above the rated capacity of the motor at full load, to insure protection against over-heating or burning out; and between the crossings of coils, at the ends, the insulation is increased with silk and oiled paper. The armature is supported on brass pedestals, by which it is magnetically insulated; and is properly balanced before mounting, to insure steadiness of motion and prevent noise when running. The shaft bushings are of brass alloy, and the bearings are grooved, and supplied with oil from a well below by rotating rings, as described on page 48.

The relative proportions of current carried by the field and armature coils respectively vary with the size of the motor, but may be illustrated by a single example, that of the 40 h. p., 500 volt motor, the armature of which is wound with No. 11 B. & S. copper wire, 3 strands in parallel, in 60 sections of 4 turns each, and carries 98.4% of the current. The two field-coils are wound with No. 20 B. & S. wire, 8000 turns to each coil, and carry 1.6% of the current.

The C. & C. Small Motors.—The construction of the C. & C. small motors, $\frac{1}{8}$ to $\frac{1}{4}$ horse-power, differs from that of the standard motor, as will be seen by Fig. 19, which represents the $\frac{1}{4}$ h. p. size, shunt wound. The field-magnets, in these small motors, are placed either directly above or below the armature, and consist of two bobbins, connected in series, and having wrought-iron cores and pole-pieces. The cores of the motor shown in Fig. 19 are bolted to an iron yoke above, and the pole-pieces to a brass base below, while in others an iron base constitutes

FIG. 19.

the yoke, from which the cores project upward, and the pole-pieces are held together by a brass plate above. There are therefore only two poles, in either case, magnetically separated, instead of the four consequent poles, without separation, found in the large motors.

The armature is supported on brass castings attached to the pole-pieces and base. It is a Gramme ring with laminated core made in two segments hinged together; and the coils, which are machine wound, are slipped on over each segment, and the two then bolted together. Hence, if a coil is injured, the ring can be opened and the injured coil removed and replaced.

These small motors are also made with series-winding,

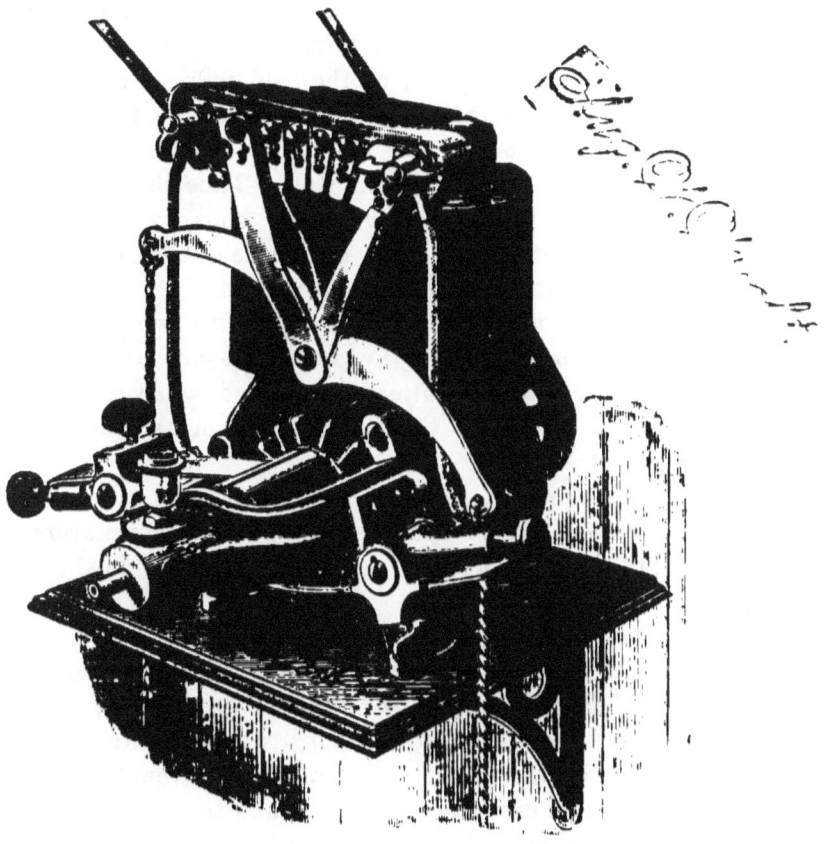

FIG. 20.

adapted to constant current circuits, and with a speed regulator, illustrated by Fig. 20, and constructed as follows: The five field-coils are connected with six stops, shown on the switch-board above, one coil between each two stops.

A jointed switch, hinged to the left stop, can be moved by a lever to the right or left, so as to make contact with any stop by its right terminal. The end stops are connected with the brushes by flexible conductors, as shown, and the current, entering by the right stop, returns to the left, after traversing the armature; and when the switch is in the position shown, passes through it and out by its left terminal, without traversing any of the field-coils. But when the switch is moved from right to left, to any stop, the current must traverse all the coils included between its right terminal and the right end stop. Thus the current may be made to traverse any number of coils, from one to the whole five, and the strength of the field, and resulting speed of the armature be varied as required.

Some of these motors are constructed with two pairs of brushes, pointing oppositely, by which the armature current may be reversed, as already described, when reversal of rotation is required. Carbon brushes, set radially, are also employed.

The C. & C. motors range in energy from $\frac{1}{8}$ to 65 horse-power, in weight, from 21 pounds to 7870, and in armature speed, from 2200 revolutions per minute, for the smallest to 750 for the largest.

The Detroit Motor.—The construction of the principal motor made by the Detroit Motor Co. will be readily understood from Fig. 21, and requires but little description. Its chief peculiarity consists in the construction of its field-magnets: these have four cores, made of wrought-iron, bolted below to a cast-iron base, of which the lower pole-piece forms a part, and above to a cast-iron yoke, which forms the upper pole-piece. Hence it has the consequent poles, already described, and as there are but two pole-pieces, each pair of cores, on opposite sides of the armature shaft, must be regarded as parts of the same magnet and not as separate magnets; so that there are but two mag-

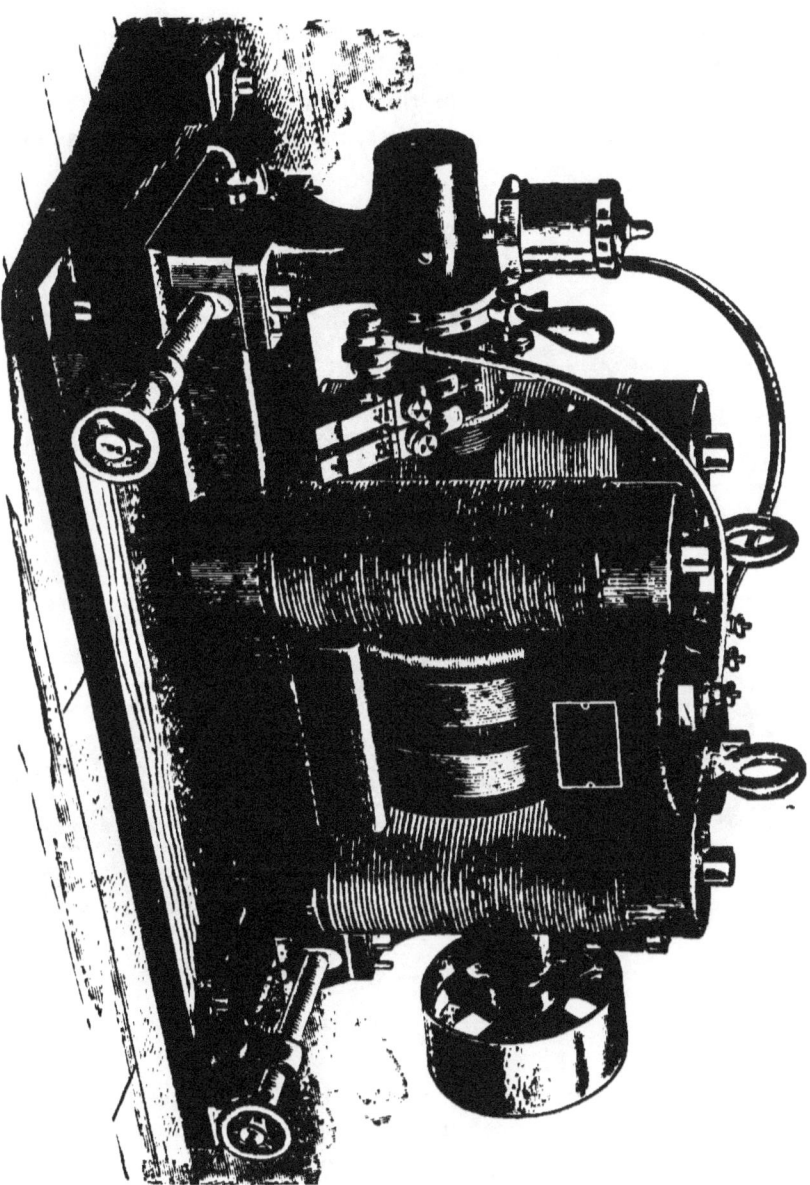

Fig. 21.

nets, each having its two opposite poles meeting at the center of each pole-piece, above and below.

By this construction the mass of iron in the cores is divided into four equal parts, and hence each core has only

FIG. 32.

half the mass which it would have in two magnets having each a single core; and the coils, being divided in like manner, are brought much closer to the cores, and the length of wire in each turn proportionally reduced, thereby proportionally reducing the resistance, increasing the electromagnetic induction, and reducing the heat liable to accumulate in interior coils.

The four coils are in series with each other, and the motor is shunt wound. Its armature is of the Gramme type, having the usual construction of core and winding found in similar armatures.

These motors range in energy from ¼ to 15 horse-power, in speed from 2400 revolutions per minute for the ¼ horse-power to 1100 for the 10 horse-power, and in weight, from 72 pounds for the former to 945 for the latter.

This company makes also motors of ⅛ horse-power, both shunt and series wound. Their field-magnets are constructed with two cores, as shown in Fig. 22, which represents the series motor, having a speed regulator, similar to the one already described.

The Eddy Motor.—This is a shunt wound motor. Its construction illustrated by Fig. 23, is very simple, and quite

FIG. 23.

different from that of the one just described. The principle assumed is, that the highest efficiency is obtained by a two-pole field-magnet, having a perfectly homogeneous structure, without projections or joints, whose poles are separated by an air space in which the armature is placed. Hence the core, yoke, and pole-pieces, as originally designed, consisted of a single, massive cylinder of soft iron, of uniform size throughout, cast in an approximately circular shape, the two pole-pieces below being separated by a circular opening sufficient to admit the armature.

But as this construction would require hand winding of the field-coils, which is much more difficult and expensive than machine winding, and liable to be less accurate, the yoke is made removable, being a separate casting bolted to the cores, as shown, and fitted to them accurately, just above the coils, so as to make the structure practically the same as if the whole were a single casting. The field-coils are wound on insulating bobbins, which are slipped on over the cores before attaching the yoke, and hence are easily removed for repairs.

The armature is of the Siemens type, the coils wound between projections on the core, which being thus brought into close proximity to the pole-pieces the magnetic resistance is reduced to the minimum, as previously shown in similar construction. The armature is supported on gun-metal yokes, bolted to the pole-pieces, as shown, which prevents any derangement of alignment, and consequent inequality of space, and hence of magnetic resistance between armature and pole-pieces, or of contact between them; a construction found also in other motors. The armature bearings are lubricated automatically by the rotating rings previously described.

The base is entirely of wood, laminated where required, to prevent warping, and furnishes perfect magnetic insulation in accordance with the principles of construction mentioned, and also electric insulation for the coil terminals and connections.

STATIONARY MOTORS.

These motors range in energy from ½ to 25 horse-power, in speed, from 2100 revolutions per minute, for the smallest to 900 for the largest, and in weight from 142 to 2200 pounds.

The Perret Motor.—The special characteristic of this motor is the lamination of its field-magnet cores, pole-pieces, and connecting yokes. These are made of thin plates of soft charcoal iron, stamped out in the required

FIG 24.

form and bolted together. It is shunt wound, and is made of three different types, distinguished respectively as A, B and C, and D; type A being of ¼ horse-power, B and C, to 2 h. p., and D, 3 to 20 h. p.; the armature in each being the Gramme ring.

Type A is a two-pole motor of the ordinary style, the magnet cores connected by a yoke above, and the pole-pieces separated by an air space below, in which the armature is mounted on supports attached to bolts through the pole-pieces, by which the plates are clamped together. In type B and C, shown in Fig. 24, two horse-shoe magnets are connected horizontally to two pole-pieces, placed above and below the armature, producing consequent poles ; and the armature is mounted on two cross-shaped supports attached to four bolts through the pole-pieces, as shown. Four of the plates shown in Fig. 25 are required to complete each set, as shown in Fig. 26, and a number of these sets to complete the structure of the core and pole-pieces.

Type D, which is the most important, is illustrated by Figs. 27 and 28; Fig. 27 being an end view, and Fig. 28 showing the construction in cross-section and the magnetic circuit. In this type we have a multipolar motor of special construction, in which the field is divided into three

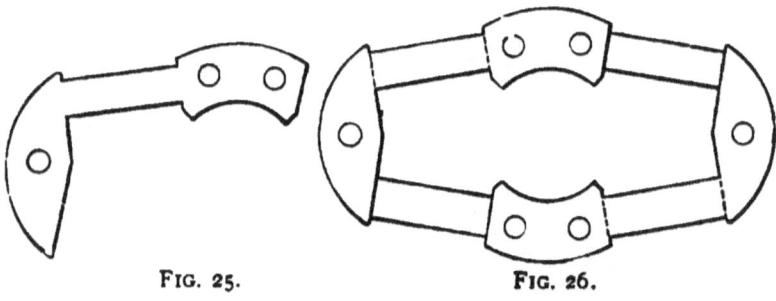

FIG. 25.   FIG. 26.

equal parts, magnetically insulated from each other, in each of which is an electromagnet terminating in two pole-pieces, giving six poles, north and south alternating, as shown in Fig. 28; the course of the magnetic circuit being as indicated by the arrows, when the rotation of the armature is opposite to that of watch hands. Each of these magnets is of precisely the same size and construction in all respects, so that the similar poles in each are of equal strength, and the distribution of magnetic force and difference of potential between opposite poles, perfectly regular in every part of the field. These magnets are supported between two iron frames to which they are attached by non-magnetic bolts passing through the pole-pieces, to which the latter are clamped by nuts and washers at a sufficient distance from each frame to insure magnetic insulation.

The armature is attached to its shaft by spiders of non-magnetic metal, and mounted on the supporting end frames

FIG. 27.

as shown in Fig. 27. Its core is laminated and insulated in the usual manner and its coils wound between the core projections shown in Fig. 28.

The armature core, field-cores, and pole-pieces being all laminated in the same plane and brought into the closest possible proximity, the magnetic resistance and formation of eddy currents is reduced to the minimum, and the shortest and most direct course furnished for the magnetic circuit; a construction which, it is claimed, gives maximum

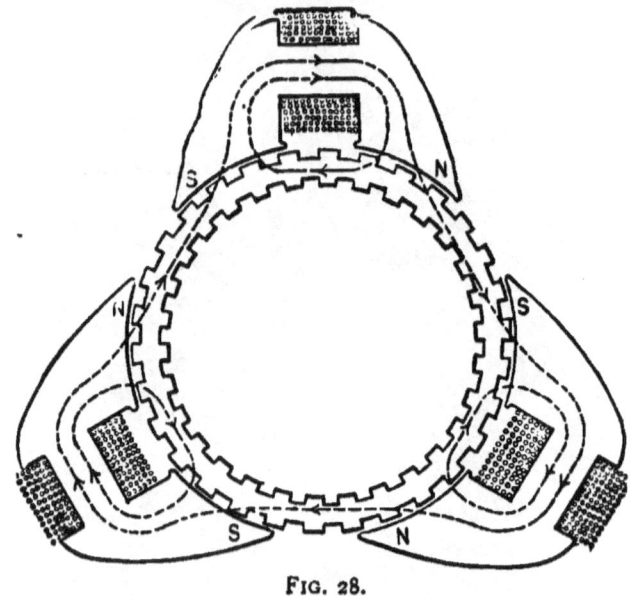

FIG. 28.

magnetism with minimum expenditure of magnetizing force.

Three pairs of brushes may be employed, and distributed round the commutator at equal distances apart, corresponding to the positions of the field-magnets; or a system of cross connections between the armature coils may be arranged by which those undergoing similar polar induction are so grouped together and connected to the commutator segments that only a single pair of brushes is required, as in a two-pole motor. In the latter arrangement copper brushes are preferred, on account of their low resistance, .

while, in the former, carbon brushes may be employed ; the higher resistance of the carbon being largely compensated by the smaller proportion of current each brush is required to carry, and the total increase of brush cross-section.

As high speed usually requires mechanical reduction to adapt motors to practical use, thereby increasing friction, expense, and mechanical complication, and wasting energy, it is highly important to reduce the motor's speed without reducing its energy. And as the development of energy depends both on the total magnetic force traversing the motor, and also on the number of times the lines of force are cut by the armature per unit of time, the latter factor may be increased, by increasing the number of poles, without increasing the motor's size, and the speed proportionally reduced with the same development of energy. This is the principle of construction adopted in multipolar motors, like the Perret, in which, it is claimed, the three pairs of poles develop the same energy at one third of the speed required in a two-pole motor of the same size. It is also claimed that the more even distribution of current in the armature, in this construction, increases its current carrying capacity, gives better radiation, less liability to heating, or injury from overloading, and permits the use of finer wire.

A sheet metal shield, fitted to the end frames and field-magnets, incloses the armature and field. The speed of the 10 h. p., 220 volt motor is 600 revolutions per minute, and its weight 900 pounds.

Alternating Current Motors.—The commercial importance which the alternating current system of electric lighting has recently assumed has created a demand for motors adapted to it, for the reasons, already given, which created a similar demand for direct current motors ; and within the last nine years various kinds of alternating current motors have been invented.

It has already been shown that when the current is simultaneously reversed in the field and armature of a motor, the direction of the armature's rotation is not reversed; and hence an alternating current will produce continuous rotation in a direct current motor, when the alternation is simultaneous in field and armature. But special construction is required to adapt a motor operated in this manner to practical work, and such construction has not yet resulted in the production of a practical motor of this kind.

The direct current motor operated by a direct current dynamo involves the double conversion of current from alternating in the dynamo's armature to direct in its field, and again from direct in the motor's field to alternating in its armature; requiring a commutator and brushes in each machine, with their wasteful resistance and sparking, and expensive construction. Hence the elimination of this double conversion by means of an alternating current motor, operated by an alternating current dynamo, has been earnestly sought; and prominent among those who have endeavored to solve this problem, have been Ferraris in Europe and Tesla in America.

In the course of his investigations, Tesla made the important discovery, that by constructing the field of a multipolar motor in a circular form, with two independent sets of coils producing successively alternate north and south poles, the polarity could be made to rotate in synchronism with the armature of a connected alternating current dynamo, and produce corresponding rotation in the armature of the motor. And, on this principle, he constructed a motor for which he applied for a patent in October 1887, which was issued May 1, 1888.

The Tesla Alternating Current Motor.—This motor, illustrated by Figs. 29 and 30, has a laminated, circular field-magnet core, with an even number of poles radiating inward, on which its coils are wound. It is constructed of thin iron plates, stamped out with the pole projections attached, and bolted together between ventilated metal caps attached to supporting end frames, as shown ; the exterior, circular part forming the connecting yoke between the poles. The coils are wound in two separate series, *B* and *C*,

FIG. 29.

whose poles alternate with each other, those of each series being wound alternately in opposite directions, so as to produce alternate north and south poles, as shown in Fig. 30.

One terminal of each series is connected to the right hand binding-post shown in Fig. 29, corresponding to T^1 in

Fig. 30, and the other terminal, in each, to a separate one of the left hand binding-posts, corresponding to T^2 and T^3.

The armature is of the ordinary Siemens type, with laminated core, wound with a few turns of comparatively small wire, forming a closed circuit, which, in the original construction, was without external connection, the two terminals being soldered together.

The mode of operation is as follows:—The motor is connected with an alternating current dynamo of special construction, having a two phase current; T^1, T^2, and T^3

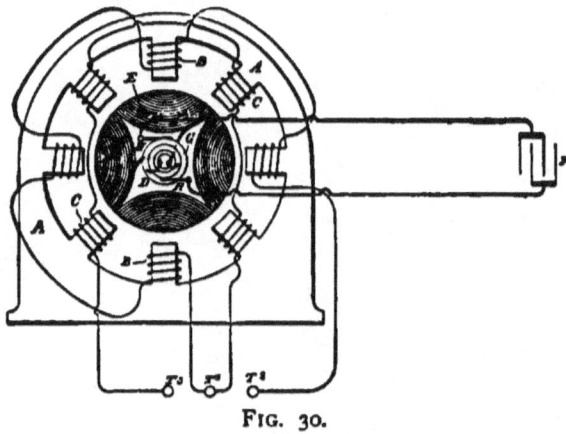

FIG. 30.

being each connected to a brush making contact with a separate collecting ring on the dynamo's armature shaft. The current entering the motor by T^1, and going to the left through field-coils B, returns to the dynamo by T^2, and, at the next impulse, goes, without reversal, to the right through coils C, returning by T^3; at the next two successive impulses, a reversed current enters, first by T^2 and then by T^3, traversing series B and C successively, and returning by T^1. The alternate north and south poles produced in the field-magnets of series B by the first impulse induce opposite, alternate poles in proximity to each in the armature; and, at the next impulse, the alternate poles in series C, acquiring the same polarity as the adjacent poles in series B acquired at the first impulse, attract the opposite poles induced in the armature by the first impulse, producing rotation. At each successive impulse, similar action occurs, so that the polarity of each field-pole, moving forward, step by step, alternately in each series, travels continuously round the stationary circular field, and is followed by an induced armature pole, occupying a relatively fixed position, producing continuous rotation of the armature ; which gives a stationary field with rotating polarity, and a rotating armature with fixed polarity. And since, in the direct current motor, the field polarity is fixed while the armature polarity has a partial rotation, Tesla prefers to call the rotating part of his motor the field-magnet, and the stationary part the armature. But as this distinction is liable to be misunderstood, it is better, perhaps, to apply the terms in the usual manner, as above. The direction of rotation is the reverse of that in the connected dynamo, as in direct current, series motors.

The successful, practical operation of such a motor depends largely on the inductive effect produced in the armature by the field-magnets, and on the difference of phase in the alternate impulses, or waves of the energizing currents in the field and the induced currents in the armature ; that is, in the relative time between the amplitudes, or highest points of successive waves. The alternate passage of the current through two separate circuits, as described, should produce a difference of a quarter phase, each successive wave beginning at the instant the previous wave had attained its amplitude, which would be sufficient, if practically attainable ; but the proper attainment of the best results, under these conditions, is seriously impaired by the self-induction of the armature circuit, by which opposing currents are generated, which partly neutralize the currents induced by the field-magnets.

To correct this, a condenser is introduced into the armature circuit, as shown at *F* in Fig. 30. The armature coils being so wound or connected that adjacent coils produce opposite poles, the two series are connected to separate collecting rings, *G*, with which brushes, *HH*, make contact and are connected respectively with the opposite plates of the condenser *F*. The two sets of plates in a condenser being insulated from each other, its well known action is to accumulate a charge of high potential on one set, by which an equal quantity of electric energy is repelled from the opposite set ; so that while there is no current through the condenser, there is a continual transfer of electric energy through the circuit in successive waves, with either an intermittent or an alternating current, and a high potential difference, or E. M. F., is maintained, which opposes and neutralizes self-induction, when the capacity of the condenser is properly adjusted to the self-induction of the coils, and periods of the currents.

To avoid the use of sliding contacts, and connection with an exterior condenser, as here shown, Tesla prefers either to place the condenser in the interior of the armature, or to use the plates of the armature core as the plates also of the condenser, making direct connection with the armature coils in either construction.

In the construction illustrated by Fig. 31, two sets of armature coils, *E* and *E'*, are joined in series, and one set is alternately undergoing maximum field induction when the other is undergoing minimum; and the points of connection, *A* and *B*, between the two, are bridged by a condenser *F*, arranged in any of the ways above mentioned. Here, the condenser, besides suppressing self-induction, strengthens the armature current in each set of coils alternately, and promotes

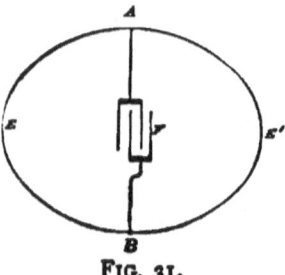

Fig. 31.

coincidence of maximum magnetic effects between field and armature poles.

In another construction illustrated, by Fig. 32, the condenser is introduced into the field circuit. This construction, in the ordinary low potential circuit, would

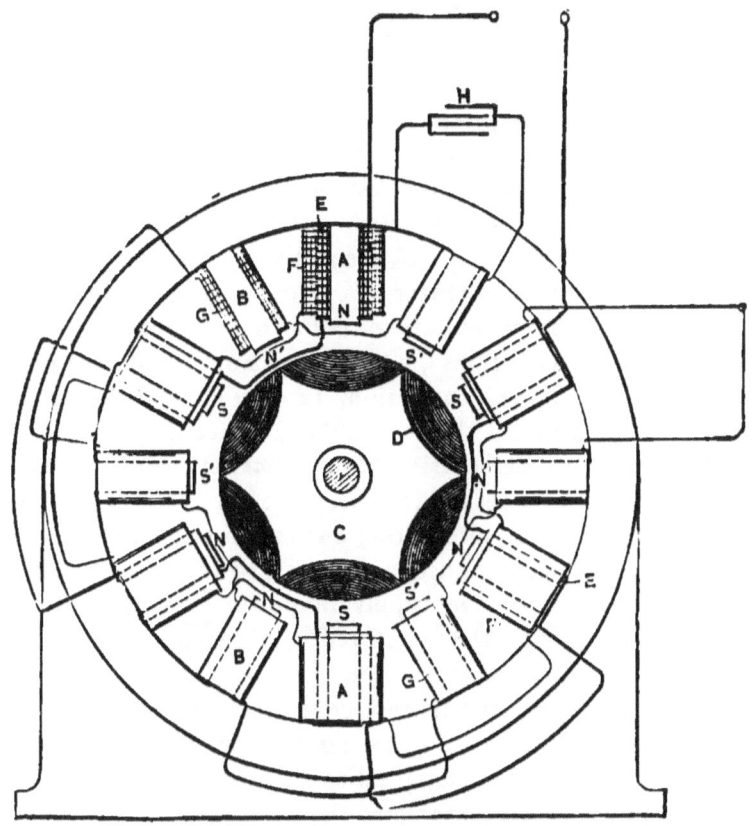

FIG. 32.

require a condenser of such size and cost, to produce the required difference of phase, as to render its use practically prohibitory; hence a secondary circuit of high potential is introduced, which is acted upon inductively by the primary circuit, and in this the condenser is placed.

The primary circuit, represented by E, is composed of coarse wire, wound on the poles A, which alternate with poles B; the winding being such as to produce the alternate north and south poles indicated by N and S. Over these coils is wound a secondary circuit F, of long fine wire, in the same direction as the primary. On the alternate poles, B, is wound a similar secondary circuit G, which is connected in series with circuit F; the direction of the winding being such that the primary currents through E shall induce alternate north and south polarity in poles B, adjacent to the similar poles A, as indicated by the letters $N'S'$. The primary circuit E is connected with the dynamo, and the secondary circuits F and G, respectively, with the opposite sets of plates in condenser F. On account of the high potential of the secondary circuits, a condenser of only small capacity is required, which must however be properly adjusted to the self-induction, rate of alternation, and potential, to produce the required difference of phase.

The Stanley - Kelly Alternating Current Motor.—This motor has been constructed with special reference to its employment on the alternating current circuits for electric lighting now in general use in America, so that such circuits may be utilized for the supply of power as well as light, instead of requiring the construction of special circuits and generators adapted to the motor.

These lighting circuits and generators are designed for current alternations of 16,000 per minute, a standard rate about double that best adapted to the operation of motors; since it proportionally increases the lagging, or false currents which impair their efficiency. Hence in the construction and equipment of new circuits it is desirable to adopt the lower rate, which will increase their efficiency for power without impairing it for lighting.

Another important point in the construction of this motor has been to give it sufficient torque, or rotary force, to overcome the friction and inertia of its armature and connected machinery, in starting, without bringing it up to synchronism with the dynamo before applying the load.

It is constructed with two circular fields, electrically independent of each other but joined together at the centre by a flanged casing, as shown in Fig. 33. They are laminated

FIG. 33.

and have interior projecting pole-pieces on which the coils are mounted, as shown in Fig. 34; and are so placed, end to end, that the pole-pieces in each come opposite the spaces between pole-pieces in the other, as shown in the two sections on the right in Fig. 35. The number of poles is dependent on the number in the connected dynamo and the speed desired. An 8-pole motor on a circuit of 16,000 alternations per minute, has a speed of 2000 revolutions per minute, since $\frac{16000}{8} = 2000$, and a drop of 7 to 8 per cent at full load.

FIG. 34.

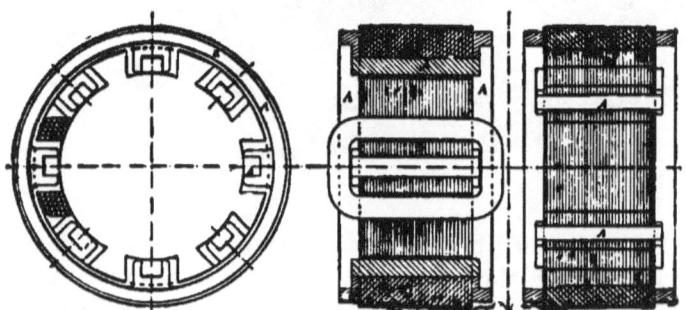

FIG. 35.

The armature is of the drum, or cylinder type with laminated core, grooved for the reception of the coils, and constructed in two sections, as shown in Fig. 36, each corresponding in length to the field section surrounding it. It is wound in a continuous closed circuit extending through both sections of the core; the coils being arranged in as many connected sections as there are poles in the field. These sections are of the same width as the pole-pieces, and are wound in two series alternating with each other; so that when the sections in one series are passing under the poles of one field, as the armature rotates, the sections of the other series are passing under the poles of the other field.

In the one and two horse-power motors, the terminals of these two series are connected together in a short circuit, without external connections; but in the larger motors one terminal of each series is connected with a separate collecting ring on the armature shaft, and the other two terminals with a third ring, the three rings being insulated from each

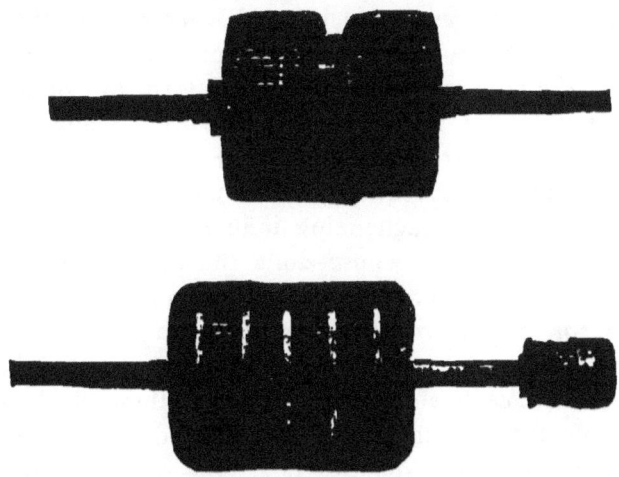

FIG. 36.

other. Three brushes, making contact with these rings, are connected by flexible conductors with a rheostat attached to the front of the motor, as shown in Fig. 33, and connected with the field coils.

The field coils are connected alternately to the opposite

sides of a three-wire circuit, and receive currents differing 90 degrees in phase; that is, each current wave appears at points in the field 90 degrees in advance of where the preceding one appeared, and is at its maximum strength when the other is at zero. These currents pass through a transformer, connected with the line, before entering the field, by which their E. M. F. is reduced from 1000 volts to 500 volts, with corresponding inverse increase of current strength.

On account of the poles in one field being in line with the spaces between poles in the other, one half of any armature coil is passing under a pole in one field while the other half is passing under a space between poles in the other field, where it is covered by the inner halves of the two adjacent field coils. As the current wave in the field coil inclosing the first pole is, at this moment, at the zero point, where the currents reverse, it produces no magnetic effect on the pole or on the armature; but the rate of change from one phase of the current to the other being then at its maximum, a current of maximum E. M. F. is induced in the armature coil, the other half of which is, at the same moment, under the magnetizing influence of the two poles in the other field, in whose coils the current wave, and hence the magnetic effect, is then at its maximum. There is therefore a pole of maximum strength induced in the armature adjacent to two poles of maximum strength in the field, one on each side of it. One of these field poles, having polarity similar to that of the adjacent armature pole, repels it, while the other, having opposite polarity, attracts it, both thus producing maximum torque in the same direction. One field and one half of the armature is therefore operating as a current inducer, while the other field and half of the armature is operating as a motor, these relations being reversed at each alternation of current.

But the current induced in the armature has a momentary lag behind the induction of the fields, so that current reversal, with reversal of polarity, is not simultaneous in armature and fields ; and this lag tends to rotate the armature alternately in opposite directions, producing a vibratory motion instead of a rotary motion, by delaying the reversal of polarity in the armature till after its poles have passed the field poles.

To correct this, a compensator is connected with each field, consisting of copper cross-bars passing through slots in the pole-pieces, and connecting two copper rings mounted on the ends of the field ring, as shown at $A\ A\ A$ in Fig. 35. As these bars are parallel to the armature coils, and have low resistance, large currents of low E. M. F. are induced in them by the armature currents, which flow alternately in directions opposite to those of the latter, dividing and returning by the rings, and neutralizing the lagging effect of the armature currents, but having no effect on the field poles ; the opposite circulation of the field currents on each side of them neutralizing field induction.

As the self-induction of the field coils produces lagging currents which would greatly reduce the efficiency of the motor, two condensers are connected with the fields by a shunt circuit, which supply neutralizing currents equal to the lagging component of the field currents. These condensers are made in sections, 8×12 inches in area and $\frac{1}{4}$ of an inch thick, each containing 50 sheets of tin foil separated alternately by special insulating material of high efficiency. Eight sections are required for a motor of one horse-power, ten for one of two horse-power, and 22 for one of five horse-power. They are contained in two small boxes, one on each side of the motor, which are connected in parallel, and each provided with a safety-fuse.

The value of the condenser has been demonstrated by practical tests, in which it was shown that the electric energy required to operate the motor without condensers is about three times that required to operate it with condensers.

The motors of one and two horse-power are started by a pull on the belt, but the larger motors, being provided with the rheostats and collectors already described, are started by moving the rheostat switch so as to cut out the resistance gradually and admit the field current to the armature; and when the resistance is all cut out, and the armature has attained full speed, its coils are short-circuited, and the field current being cut off from them, the armature current is then generated by induction on a closed circuit, without external connection, as in the smaller motors. The resistance coils in the rheostat consist of spirally wound iron ribbon.

This motor is provided with self-oiling ball bearings, and will run continuously with but little care except an occasional supply of oil. It has been fully tested in actual service and found to operate satisfactorily. It is made in the various sizes required for the operation of either light or heavy machinery, those ranging from 1 to 15 horse-power being most in demand. As an example of its weight in proportion to its power, it may be stated that the five horse-power motor weighs 750 pounds.

Single Phase Alternating Current Motors.—Two phase alternating current motors, of any construction, require the addition of a third wire to the two-wire lighting circuits in common use in America and Europe, and also the rewinding of the generator's armature in two circuits so related to each other as to generate currents having a difference of a quarter phase, or 90 degrees; hence single phase motors fully adapted to these two-wire circuits, as now constructed and equipped, have been earnestly sought for, and many such are now in actual use, prominent among which are the Dahl and Wagner in America, and the Brown and Oerlikon in Europe. Each of these has already attained a limited commercial development, especially for the smaller sizes which do not require powerful starting torque, but none of them are yet sufficiently perfected and developed to permit a full detailed description, or to meet the demand for motors of large horse-power; though the rapid progress recently made indicates that stationary motors of this class, in sizes up to 100 horse-power, will soon be produced, and also railway motors of 20 horse-power, surpassing the direct current railway motors in efficiency. All these motors have circular fields, more or less similar to those already described, and armatures chiefly of the drum type.

The low rate of current alternation employed on European alternating current circuits is much more favorable to the employment of motors of this class than the high rate on similar American circuits.

The Brown Single Phase Alternating Current Motor.— Some of the principal points in the construction of this motor can be given. The coils in both the field and armature are imbedded in laminated cores, either in holes close to the inner circumference of the field core and outer circumference of the armature core, or in grooves on the surface of each, closed by a superimposed winding of iron wire, or nearly closed by side projections on the teeth; thus practically eliminating core projections with air spaces between them, or spaces filled with copper wire, from both field and armature. The object of this is to reduce the magnetic resistance of the air gap to its minimum, and thus avoid the necessity of employing an abnormal magnetizing current, with its waste of energy and injurious heating.

The field is of the multipolar type; and the armature of the drum type, wound in sections, in a closed circuit; its coils being short-circuited by soldering their terminals to two copper rings, one at each end.

Starting torque may be obtained by a second field-winding connected with the first in parallel, but of relatively smaller section, wound in the spaces between the coils of the first winding; one set of coils having a drum winding and the other set a ring winding, giving the latter greater self-induction than the former. This difference of self-induction causes a difference of phase between the two windings which produces rotation of the poles in the stationary field, with starting torque. After the required speed has been attained, reversal of polarity approaching synchronism with that of the connected dynamo, the supply of current may be cut off from the second winding.

This method is similar to one devised and patented by Tesla in 1889. Difference of phase may also be produced by suitably proportioned transformers.

Starting torque may also be obtained in the same manner as in the Stanley-Kelly motor, and resistance for regulation of starting current be employed in a similar manner. A condenser, composed of conical iron cells, insulated from each other and filled with a solution of soda in water, may also be introduced into the field circuit for the same purpose.

The larger motors of this construction are similar in external appearance to the Stanley-Kelly motor.

CHAPTER IV.

APPLICATIONS OF THE STATIONARY MOTOR.

General Remarks.—The convenience and economy of the electric transmission and application of power are often very great, especially as it can thus be distributed to points remote from its source more economically often than by any other means. In a large factory, operated by steam or water power, the transmission of power by electric wires may often be found preferable to its transmission by shafts and belts. In cities power can be generated at a central station and distributed over a large area for operating various kinds of machinery, as printing presses, lathes, sewing machines, elevators, and pipe organs, at points where the employment of engines would not be convenient or economical. Water power may, in this way, be transmitted from falls in a rural district where it is serving no useful purpose, and distributed for industrial purposes in a distant city.

The distribution of electric energy in this way, whether generated by steam or water power, opens a wide field for the introduction of the motor; and as the maximum demand for motor work occurs during the day, and that for electric lighting during the night, these two kinds of work can be performed most successfully in conjunction, so that the maximum constancy of work can be obtained from the engines and dynamos at the most economic rates, and the loss incident to the cooling and heating of furnaces and boilers, and the idleness of capital invested in expensive machinery be reduced to the minimum.

It becomes evident from this that the numerous uses to which the stationary motor is now applied must be largely multiplied in the near future, as the number of central distributing stations for light and power in towns and cities is increased and the areas of those now established extended ; and that the motor will not only do work now performed by manual labor in small factories and elsewhere, but will replace water motors, caloric engines, and small steam and gas engines ; requiring less attention, occupying less space, incurring less risk, operating more quietly, and without the annoyance incident to heat, smoke, dirt, escaping steam or gas, dripping water, supply of fuel, and removal of ashes, cinders, and debris. Incident to this will be the increased demand for large engines of the most approved construction to operate the dynamos, the larger development of numerous industries requiring small power, and the greater convenience and facility with which power may be applied in large factories.

The motor is especially well adapted to mining work, since power, generated at the mouth of the shaft and electrically conveyed, may thus be applied to machinery for drilling, pumping, ventilating, and running tram cars in the interior, where the fire, smoke, and steam make the use of the engine prohibitive. Water-power may also thus be made available, when found in the vicinity of mines located where scarcity of fuel or its transportation into mountainous regions, difficult of access, makes the use of steam power too expensive. Power may be electrically conveyed by wire over rugged ground and into the interior of the mine, not only far more economically than by any other method, and the apparatus put in working order more expeditiously, but its maintenance in proper repair is far less expensive.

This is especially true of the electric system as compared with the pneumatic system in common use in mines. The flexible wire or cable conductors can easily traverse the intricate, low, narrow, winding passages of the mine, with their numerous grades and levels, where the location and construction of the stiff pneumatic tubing is difficult and expensive. The removal of electric apparatus and conductors from one part of the mine to another, as often required, is effected expeditiously and inexpensively, while

FIG. 37.

the similar removal of pneumatic tubes and apparatus is slow, difficult, and costly. The repair of an electric wire, broken by the accidental fall of rock or mineral, is the work of only a few minutes, and the temporary interruption of slight importance, while the repair of a pneumatic tube causes serious delay, annoyance, and expense. The loss of power in transmission, by leakage or otherwise, is also far greater in the pneumatic than in the electric system.

The methods by which the motor may be adapted to various kinds of work will be best understood by giving illustrated descriptions of a few of its numerous applications.

Electric Fans and Ventilators.—The operation of a rotating fan, one of the most common applications of small motors, will be readily understood from Fig. 37, which shows the fan mounted on the armature shaft. The motor may be placed in any convenient position on a desk or table, where connection with an electric circuit can be made, occupying no more room than a table lamp, and will run continuously, at any speed desired, and furnish an air current, as required, with no other attention than occasional renewal of the lubricating oil. The electric energy required to operate it is about equal to that for a 16 candle-power, incandescent lamp; and where such lamps are already in use, one may be removed and the motor connected in its place. For special, temporary use, where such connection may not be available, and trouble and expense is of less importance than proper ventilation, as in a sick room at a private residence, current may be supplied by a suitable primary battery, or storage battery, as may be found most convenient.

Ceiling fans of different kinds, adapted to hotels, restaurants, and factories, may be operated by the motor, either directly, or by belt and shaft connection, as may be found most convenient.

The application of the motor to a blower, for ventilating ships and interior rooms in large buildings, is illustrated by Fig. 38, which shows one of the larger motors attached to the blower; the rotating fan inside, being mounted on the armature shaft. Blowers, operated in this manner, are in use on several of the largest U. S. cruisers.

Electric Operation of Pipe Organs.—In Fig. 39, copied from a photograph, the operation of the bellows of a pipe organ by a motor in actual use, is shown in detail. In this

APPLICATIONS OF THE STATIONARY MOTOR. 85

case the application of power requires special, automatic adjustment to secure an even alternate movement of the bellows, and a regular supply of current corresponding to

FIG. 38.

the power required. The motor is shown mounted on a shelf at the left, its pulley connected with shafts and belting by which a double reduction of speed is obtained, and these with a crank connected to a vertical bar, hinged to the lever by which the bellows is operated.

The electric circuit is connected, as shown, with a rheostat mounted on a shelf above the bellows, and the rheostat switch connected by a vertical bar to the lower part of the bellows. The motor power is required to inflate the bellows, which, being weighted above, is compressed and the

86 *THE ELECTRIC TRANSFORMATION OF POWER.*

air expelled by the downward pressure of the weight, during which the motor power must be reduced; and this alter-

FIG. 30.

nate increase and decrease of power is regulated by the rheostat. When the bellows is inflated, as shown, the switch is on the uppermost contact, and the current traversing the maximum resistance; hence its strength, and the supply of power, is at the minimum; as the bellows is compressed, the switch is moved downward, the resistance reduced, and the current strength and supply of power correspondingly increased till the lowest contact is reached, when the maximum current strength and supply of power for reinflation is attained. By this method of regulation, smoothness, and evenness of mechanical movement, so important in the blowing of an organ, is obtained, and injury to the armature by a sudden influx of current in excess of speed, prevented.

Another method of regulation, entirely mechanical, is to allow the motor to run at a constant speed, and by the automatic movement of a belt-shifter, connected with the bellows, shift the belt to a loose pulley and back again to the motor pulley alternately as the bellows moves up and down; thus withdrawing and renewing the power alternately as required.

Each organ requires special adaptation of the motor to its construction, size, and the conditions under which it is operated; but the facility with which the electric motor can be adapted to these varying conditions commends it for this use above every other motor, wherever a supply of electric current is available. It has been successfully applied to organs of every size and description, whether originally designed to be blown by manual power or a mechanical motor.

The various mechanical motors employed for this purpose are the steam engine, gas engine, naphtha, hot air, and caloric engines, and the water motor. Of these the water motor and steam engine have been the most extensively employed; the former, limited to localities where connection with water works renders its use practicable, being the least objectionable. The principal difficulties attending its application are deficiency or excess of pressure, the latter producing unevenness of mechanical motion, clogging of the motor valves by foreign matter, and liability, in winter, to freezing in the supply pipes. The application of the various engines mentioned is costly, and their use requires the constant care of an engineer. In warm weather their heat is always an objection, and the generation of steam, even in steam heated churches, expensive; besides the noise, which in the gas engine especially is difficult to control, often makes their employment in churches prohibitive; while the odor from the latter engine is liable to penetrate into the audience room and vitiate the air.

The electric motor is free from all these objections, runs noiselessly, occupies but little room, can be placed on a shelf, or in any convenient spot out of the way, is directly under the control of the organist and requires no separate attendance, and can be so regulated as to furnish instantly the wind supply required, either for the full power of the organ or the most delicate shading of the music, with perfect evenness, thereby giving a quality of tone unequalled by any other means of blowing.

The size of motor required for church organs ranges from 1 to 3 horse-power, and for small organs suitable for private residences, from $\frac{1}{8}$ to $\frac{1}{2}$ horse-power.

Electric Elevators.—None of the different methods of operating elevators is so satisfactory and economical as that by the electric motor, when proper construction is employed. The principal requirements for a passenger elevator are smooth, rapid movement, easy starting and stoppage, without jolt, and perfect security against accident; all of which can be obtained by means of the electric motor.

As an electric motor consumes power only when running, and always in proportion to load, an elevator operated in this manner by current supplied from a central station and paid for by the number of watts consumed has a great economic advantage over a steam or a hydraulic elevator; the steam elevator requiring the service of an engineer and consuming fuel not only while running but during the time between trips when it is idle, which often exceeds the average time it is in operation, and the hydraulic elevator costing as much to operate it without load as with maximum load; while the electric elevator requires only the service of the attendant who runs it, and who need not be an electric expert; the motor requiring little attention except to renew the supply of oil to the self-oiling bearings, and the occasional renewal of worn-out brushes.

When power is applied by belts and shafting, as is sometimes done with freight elevators, mechanical reversal of rotation may be obtained by a crossed belt, loose pulley, and belt shifter operated by a cable running through the car, instead of reversing the rotation of the armature, as is usually done with passenger elevators.

Twenty horse-power is the size of motor most commonly required for a passenger elevator carrying an average load of two and half persons, and making four or five hundred trips a day.

The Otis Electric Elevator.—This elevator is illustrated by Fig. 40, which shows the operating apparatus in perspective. It is operated by the Eickemeyer motor, shown on the left, which is a compound wound machine with a Siemens armature, the shaft of which is connected at the brake wheel A, by an insulated coupling with the shaft which rotates the drum B. On the latter shaft is a worm gear, adapted to a large gear wheel C, as shown in Fig. 41, in which two such wheels, employed in larger elevators, are shown, having right and left gears, which mesh with each other and with two worm gears on the shaft. When the armature rotates, it rotates these wheels by means of the worm gearing, turning them toward or from each other, according to the direction of the rotation, and thus neutral-

90   THE ELECTRIC TRANSFORMATION OF POWER.

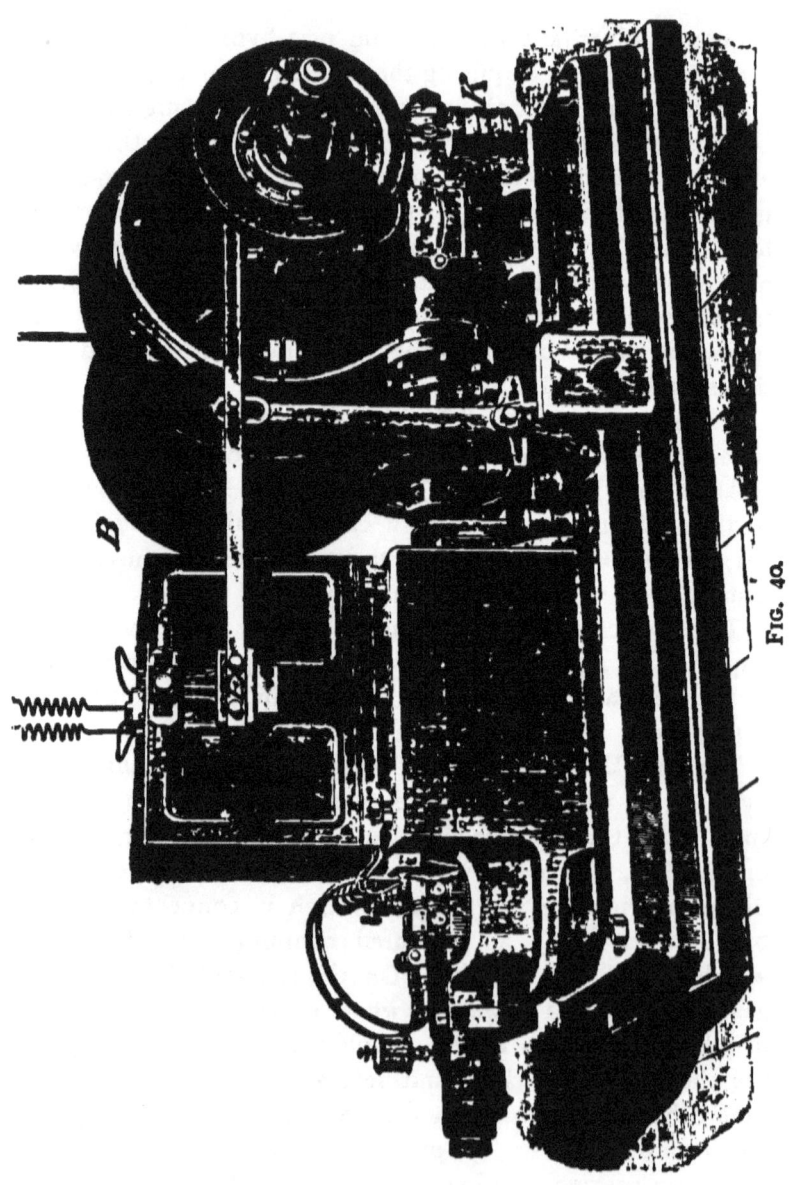

Fig. 40.

APPLICATIONS OF THE STATIONARY MOTOR. 91

izing the end thrust on the shaft, which, in large elevators, produces excessive friction and heating; it being taken up by the worm gears, as they are thrust toward or from each other.

The gear wheel C rotates the cast-iron drum B, being mounted on the same axle. On this drum are wound in grooves, four wire cables which pass over stationary pulleys at the top of the elevator shaft. Two of these cables are attached to the top of the car and employed to hoist the load, and the other two, which are wound in the opposite direction, support a counterpoise equal to half the maximum load; the hoisting cables winding on the drum, as the car ascends, and the counterpoise cables winding off, this process being reversed during the descent; so that each pair occupies the grooves vacated by the other pair,

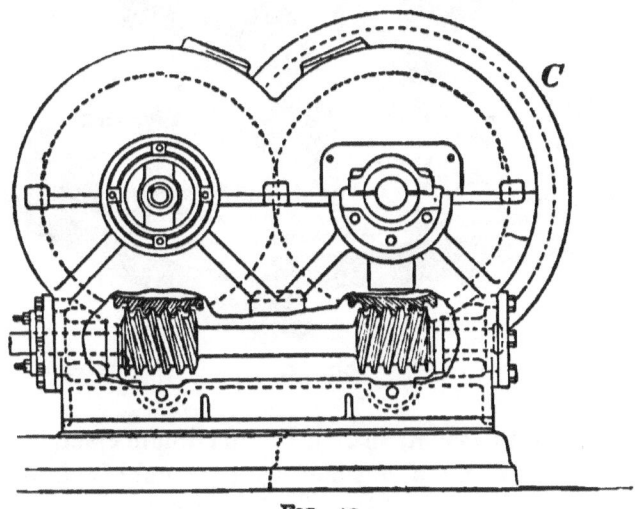

FIG. 41.

being attached to opposite ends of the drum. There are besides these, two other cables attached to the top of the car, which pass over pulleys above and support a counterpoise nearly equal to the car's weight; sufficient difference being allowed to insure tension of the cables when the car descends empty; both counterpoises traveling in the same guides.

Half the maximum passenger load being counterpoised on the drum, power to this amount is stored up by the hoisting of this counterpoise as the car descends, and this power is expended in helping to hoist the load as the car ascends; the work being thus approximately equalized in

FIG. 42.

the ascent and descent, so that the maximum electric strain on the motor is only about half what it would be if it were required to hoist the full load during ascent.

Case D, above the motor, incloses the rheostat, which with its reversing switch and various connected apparatus is shown in Fig. 42. This reversing switch is a wheel which carries four brush contacts on its rim, and is rotated alternately in opposite directions by a lever mounted on its shaft and operated by a spring locking device connected at *E*, Fig. 40, with the shipper bar which extends to the shaft of the gear wheel *C*, and terminates in a rack which meshes with a pinion on a loose pulley *F*, mounted on the end of this shaft. This pulley is connected by a system of belts, wheels and pulleys with a wheel or lever in the car, by the movement of which the reversing switch is operated through the connections described, and the armature current reversed, as required for the alternate reversal of rotation which produces the ascent or descent of the car.

The four contacts referred to consist of two long conducting segments, shown on the upper and lower parts of the rim, and two short insulating segments, shown at the sides, by which the long ones are insulated from each other. Four brushes are shown which make contact with these segments, two above and two below. When the wheel is rotated to a position in which the four brushes rest on the four contacts, the brushes are all insulated from each other, and no current can traverse the motor ; hence this is called the neutral position : but when rotated to a position in which the two upper brushes rest on one conducting segment and the two lower on the other, current entering by one of the upper brushes will flow through the segment to the other, passing thence to the armature by the connections shown, and, having traversed it, will return to one of the lower brushes, and through the segment to the other brush.

But when the wheel is rotated to a position in which the conducting segments are at the sides and the insulating segments at the top and bottom, current entering by the same brush as before will pass through the segment to the adjacent lower brush and thence to the armature, and hence its direction through the armature will be reversed, and it will return to the opposite lower brush, and through the segment to the upper one.

The armature current passes from the reversing switch through a solenoid coil in the rear, at the left, with which the series field coils are also connected, going thence to the upper end of the series of contacts shown at G, on the right, and traversing the resistance coils connected with these contacts, when the brush H is in the position shown, returning to the dynamo through the wire connected with the lower end of this brush.

A soft iron armature, connected with this brush by a hinged lever, has a free vertical movement within the solenoid, into which it is attracted in proportion to the strength of the current, pulling the brush H downward in opposition to a weight I, which tends to move it upward; the upward movement reducing resistance and the downward increasing it. Hence as the current strength increases the increased resistance tends to reduce it, this process being reversed as it decreases, and in this way the current is automatically adjusted to the work required of the motor, and is not controlled or regulated by the elevator boy, whose only business is to move the wheel or lever to the position for starting, stopping, or reversal.

At the lower left corner is shown a snap switch with V-shaped contacts on each side, by which the main circuit from the dynamo is closed or opened for the admission or exclusion of current. When the reversing switch is in the neutral position, the snap switch is vertical, as shown, and the circuit open, but when the reversing switch is rotated to either of the other positions, a cam connected with it turns the snap switch diagonally to the right or left, according to the direction of the rotation, and closes the circuit.

The locking device, which is controlled by springs which regulate its movement, as shown above E, Fig. 40, holds the reversing switch in place when in the neutral position; its resistance being overcome by the movement of the shipper bar in either direction.

The brake at *A* is operated by a vertical lever connected above with the shipper bar, as shown, and terminating below in a short horizontal cross-bar which acts as a fulcrum to lift the brake off its wheel in opposition to a weight *J*, as the lever is moved to the right or left. When the reversing switch is in the neutral position this lever is vertical and the brake in contact with its wheel on the shaft; but when the reversing switch is rotated in either direction by the horizontal movement of the shipper bar, admitting current to the motor circuit, the simultaneous movement of the lever releases the brake, permitting rotation of the shaft.

In the rear of the drum is a horizontal bar hinged by two arms on opposite ends of the drum shaft, and supported by the tension of the hoisting cables. A lever connected with one of the arms operates a clutch consisting of a pair of toothed wheels, one attached to the loose pulley *F*, and the other to a collar which slides on the drum shaft. If the cables should become slack by any accident, as the stoppage of a descending car by some obstruction while the cables continue to unwind, they will drop away from this bar, which will then be pulled down by a weight *K*, and, by means of the connected lever, will force the wheel attached to the sliding collar against the other, locking the clutch, which, by its connection with the loose pulley, will move the shipper bar, bringing the reversing switch to the neutral position and applying the brake.

On the outer end of the drum shaft is a clutch composed of three separate parts, inclosed by the hood *L* which is attached to the loose pulley; two of these parts are attached to the shaft, and the other is a nut which travels between them on a screw on the shaft, moving to the right or left in guides connected with the hood, as the drum rotates in either direction, the distance being adjustable by attaching the outer part at any required point. This

96   THE ELECTRIC TRANSFORMATION OF POWER.

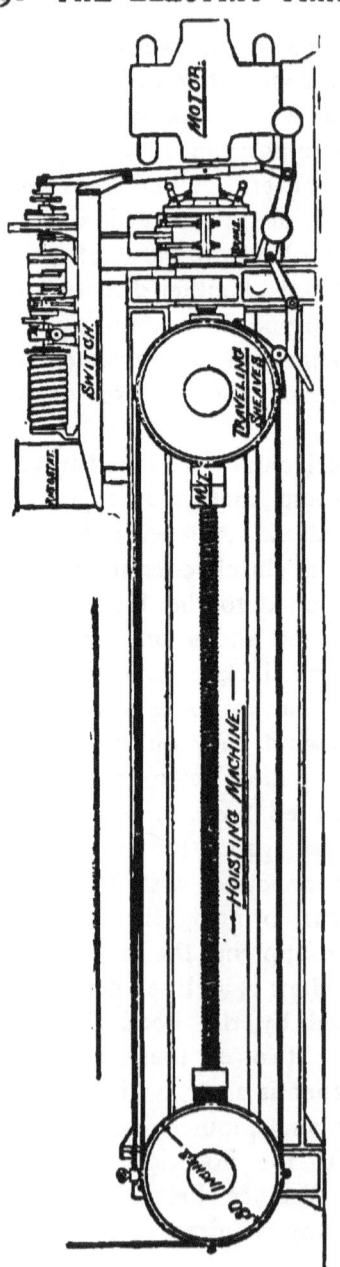

adjustment is such that the traveling nut reaches the limit of its range in one direction when the car arrives at the top of the elevator shaft, and in the other direction when the car arrives at the bottom, and in each case locks with one of the stationary parts, and, by the loose pulley connections, brings the reversing switch to the neutral position and applies the brake as before, stopping the rotation.

These elevators have been in successful operation since 1890, giving satisfactory service.

The Sprague-Pratt Electric Elevator. — This elevator is especially designed for service in high office buildings where speed and safety are the most important requisites. Its general construction, which is mechanically similar to that of

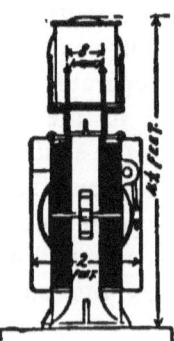

hydraulic elevators, is shown in the sectional diagram in Fig. 43.

Two pairs of steel multiple sheave blocks, each 30 inches in diameter and carrying a number of sheaves, are mounted on a horizontal steel frame having upper and lower guides for their support, securely attached to end pieces; and around these sheaves passes a steel hoisting rope, extending thence to the top of the building and over pulleys down to its attachment to the top of the car.

The left pair of blocks, shown in cross-section in Fig. 44, are stationary, and the right pair, similarly constructed, are mounted on trunnions on each side of a cross-head which travels on a central screw rotated by the motor shown on the right in Fig. 43.

As the traveling sheaves are moved away from the stationary sheaves by the action of the screw, the car ascends with a speed proportionate to the number of multiplying sheaves. If there are 8 such sheaves, it will ascend 8 feet for every foot of horizontal travel of the sheaves. Hence 25 feet of horizontal travel would elevate it to the top of a building 200 feet high. But the force required to produce this relative speed must be 8 times as great as that represented by the load elevated, plus the loss by friction and otherwise. Hence, allowing 50 per cent for this loss, which is about the proper allowance for this elevator, a force of 6 tons would be required to elevate a load of 1 ton.

At the center of the traveling cross-head is a compound nut, consisting of safety and hoisting parts, the former inoperative except for safety, the latter held against the cross-head by friction, which prevents its rotation during the movement of the car, so as to move the cross-head as the screw rotates. The screw passes through this nut and the cross-head and terminates between the end of the frame and the motor in a brake wheel, above which is supported a steel and leather brake which acts on the wheel through a double toggle joint controlled by a weighted lever operated by an electromagnet through the agency of a small controlling motor mounted above the frame. This brake is not employed to vary the speed, but to lock the screw when at rest, and to assist in stopping it in case of accident.

The controlling motor is operated by a switch in the car through a secondary switch shown in Fig. 43, and it operates not only the brake lever but also the rheostat switch by which the current of the hoisting motor is controlled. This switch, which is of a peculiar spiral construction, is shown between the controlling motor and the rheostat.

The connection of the controlling motor with the car switch is made by a flexible cable inclosing four copper wires, which is looped to the car from a midway point in the elevator shaft. And as all the movements are under control of this motor, the elevator man cannot, by carelessness or otherwise interfere with them ; being only required to move the car switch to the "up" position for ascending, the "down" position for descending, and the central or "stop" position for stopping ; the operation of the rheostat switch and the application and release of the brake being automatically controlled by this motor.

The armature shaft of the hoisting motor is in line with the extension of the screw, and connected with it by a sliding coupling ; and on this shaft is also mounted the centrifugal governor which operates the brake when the speed exceeds the safety limit.

It is evident that whether the traveling sheaves are moving from the stationary sheaves or toward them, the strain on the rope, and hence on the screw, must always be toward the stationary sheaves. This strain is sustained at the motor end of the screw by a thrust bearing against the end of the frame.

The friction in the thrust bearing, hoisting nut bearing,

and sheave bearings, is reduced by half inch steel balls which change the sliding friction to a rolling friction. In the hoisting nut there are 300 of these balls, which travel continuously in the 12 convolutions of the nut thread, between it and the screw thread, so that no part of the nut touches the screw. And as each ball circulates through the nut, and emerges from it at one end, it is pushed back through a tube by the balls behind it, and re-enters the nut at the other end, 50 to 75 balls per second passing a given point. The balls circulate in a similar manner in the other bearings, except that no return tube is required. As a strain of 20,000 to 40,000 pounds is required to crush one of these balls, and the maximum working strain on each does not exceed 125 pounds, absolute security against breakage is assured.

The reduction of friction in this manner is a very essential point in the construction of this elevator, and it is claimed that the difficulties usually found in ball bearing construction have been entirely overcome, and that about 50 per cent of the electric energy supplied to the motor is utilized in lifting the load, which is about double the efficiency of similar hydraulic elevators.

The hoisting motor has a light field, separately excited, and variable at will. The rheostat, through which current is supplied to it, is constructed with a large number of resistance coils, so as to reduce the potential difference between successive contacts; and these coils are connected with bronze contact plates of large area and ample carrying capacity, which are arranged in a spiral form, as already mentioned, on the outside of a hollow cylinder, and connected with the resistance coils in the rheostat by wires passing through the inside of the cylinder; and over these contacts, which occupy more than ten feet of lineal space, a carbon brush, through which the current passes to the hoisting motor, travels at a high rate of speed.

This rheostat will carry the current with the brush in any position, including any portion of the resistance, for any length of time, without injurious heating of the resistance coils or injury to the brush or contacts.

The ascent is produced by moving the car switch to the "up" position, thus closing the circuit of the controlling motor, which, having first closed the circuit of the hoisting motor through its armature, and by the operation of the rheostat switch, cut out sufficient resistance to prevent the car load from reversing the rotation of the hoisting motor's armature, releases the brake and continues to cut out resistance until there is sufficient current admitted to start the hoisting motor and produce the required speed. All of which is done in the minimum time required to attain full speed by a gradual acceleration without discomfort to the passengers.

The ascent is stopped in a similar gradual manner by moving the car switch back to the central position, which reverses the operation of the controlling motor, introducing resistance into the circuit of the hoisting motor, applying the brake and opening the circuit.

When the car has reached its uppermost limit, it is brought to a gradual stop by a trip operated by the traveling cross-head, which applies the brake and cuts out the hoisting motor's current. At the same instant, the hoisting nut, which, as already explained, is kept from revolving, during the ascent, by the friction produced by the strain on the rope, is brought into contact with a collar at the end of the screw, which locks with it, causing the nut to revolve with the screw without stopping the rotation of the armature or straining anything, but effectually stopping the car, without preventing its being lowered again by the movement of the car switch.

By stopping the car switch at any intermediate point between the "up" and central positions, the car will ascend at the diminished speed corresponding to the resistance in the circuit.

In case of accidental interruption of the current, an electromagnet in the shunt field applies the brake and, at the same instant, returns the main switch to the neutral position, thus preventing injury to the electric apparatus by a sudden resumption of the current with the resistance cut out. But, in ascending, this action does not take place till the car has stopped, from the fact that the shunt fields being separately excited and their terminals joined to the same line wires that supply the armature of the hoisting motor, when the line current fails, or the main fuses melt from an overload, the fields are left in series with the armature, which continues to revolve by its own momentum, and thus generates sufficient current through the electromagnet to prevent the immediate application of the brake; thus avoiding the liability of the ropes slacking by a sudden stop with the car at full speed upward. The time of this action can be varied to any degree required.

The descent is produced by moving the car switch to the "down" position, causing the controlling motor to begin to release the brake; the hoisting motor's armature terminals being short-circuited, the resistance cut out, and the fields separately excited, the car, with its load, begins to descend by its weight, and in doing so moves the traveling sheaves toward the stationary sheaves, causing the hoisting nut to reverse the rotation of the screw, thereby reversing the rotation of the hoisting motor's armature and causing it to generate current as a dynamo; thus producing a resistance which limits the movement of the car to a very slow degree of speed. The movement of the controlling switch being continued automatically by its motor, the brake is fully released, and more resistance introduced into the circuit of the hoisting motor, reducing the resisting motor current and permitting the car to descend more rapidly. This acceleration of car speed accelerates the rotation of the armature, increasing the resisting current; and thus, by a proper adjustment between this current resistance and the rheostat resistance, any required speed in starting, stopping, and descending can be obtained; the control being so perfect that movements of $\frac{1}{4}$ of an inch can be readily made, and the speed varied from 6 inches per minute to 400 feet per minute.

The car may be stopped at any point in its descent by moving the car switch back to the central position, which simply reverses the operation of starting down.

In case of accidental interruption of the current, the circuit of the electromagnet which controls the brake is fully opened, and the brake applied at once without strain on the car.

To produce an easy, absolute, automatic stop at bottom, there is placed at the elevator end of the screw, a spring buffer composed of elastic rubber washers and thin steel disks, which acts as a brake when the hoisting nut meets it, reducing the speed of the screw, and, if the pressure is sufficient causing the nut to lock and revolve with the screw after compressing the buffer about $1\frac{1}{2}$ inches, or one foot travel of the car; and thus as at the termination of the ascent, stopping the car without stopping the motor.

There is also arranged for the same purpose, a series of variable electric resistances connected with bronze contact plates on the traveling cross-head, over which a carbon brush connected with the electric circuit moves. The first contact is made when the car is about 6 feet from the bottom and the resistance gradually cut out, increasing the resisting motor current, till the armature is short-circuited, and the speed so reduced that the nut when stopped compresses the buffer only about $\frac{1}{2}$ an inch, or about 4 inches travel of the car, making the softest possible stop, even at a speed of 400 feet per minute.

As an illustration of the effectiveness of these two devices, a loaded car in a new elevator, in which the other safety devices were incomplete and inoperative, descended freely from the fifth story, at a speed of about 800 feet per minute, and was stopped at bottom, in this manner, in a space of about 18 inches travel, without injury to the car or apparatus, or shock to the passengers.

In addition to the other safety devices is the safety nut already mentioned as connected with the hoisting nut. It has a deeper thread than that of the hoisting nut and no ball bearing, and this thread is normally out of contact with the thread of the screw, but in case of accident to the hoisting nut which would break its hold on the screw's thread, the safety nut would immediately interlock firmly with the screw, and rotating with it, stop the movement of the traveling sheaves.

There is also the centrifugal governor referred to, mounted on the armature shaft of the hoisting motor, where it is easy of access, and so adjusted that it operates a lever which applies the brake and stops the car, when the speed exceeds the safety limit.

The screw is made of special steel forged under tension and torsion strains, and capable of sustaining twenty times the greatest strain normally applied to it. The hoisting nut is also made of steel hardened and finished by a special process, and it has been found to show scarcely any wear after a year's test in specially onerous service.

The rope is secured at the ends in solid metal fasteners, each constructed with a double spiral groove in which about three feet of rope is so wound and looped as to resist any strain to which it may be subjected.

The whole construction of this elevator is such as to insure a perfectly smooth and easy movement in stopping, starting, and running at any rate of speed from 250 feet to 450 feet per minute. It is giving satisfactory service in the highest office buildings, carrying loads of 1800 to 2400 pounds, 10 to 24 hours a day; and is operated with one third of the coal consumed in similar hydraulic elevators.

Electric Dock Hoists.—The electric motor is applied to the operation of dock and ship hoists in various ways, requiring different methods of construction. In such hoists it is often found convenient to combine a hauling winch and a hoisting drum in the same machine; the former being employed to draw heavy loads to the hoist, and the latter to elevate them. A hoist of this construction is illustrated by Figs. 45, 46 and 47, a side elevation being shown in Fig. 45, an end elevation in Fig. 46, and a cross-section of the latter in Fig. 47.

In Fig. 45 is shown the motor A, from which extends the steel shaft C, which carries a steel worm gearing which engages a bronze gear wheel on the shaft of the drum, inclosed in the iron case K, shown in Figs. 46 and 47; the gearing being shown in section at J, Fig. 47, and the gear wheel at D, where it engages the gearing. An iron hoisting drum I and hauling winch-head G are shown mounted on the steel shaft F. The drum is loose and can be moved horizontally through the space shown near the center of the shaft. In its interior is an iron friction clutch E, shown in section above and below the shaft. At the right end of this shaft is shown an iron wheel H, with handles projecting inward for safety; a screw cut on the inside of its hub engages a screw on the end of the shaft by which the drum can be moved against the friction clutch or withdrawn from it; the wheel being connected to the drum by a collar, as shown, with interior projections running in grooves on the hub of each.

The rheostat is placed under the hoist at P, and connected with the controller M, inclosed for safety in the iron case O, from which projects the operating lever N,

APPLICATIONS OF THE STATIONARY MOTOR. 105

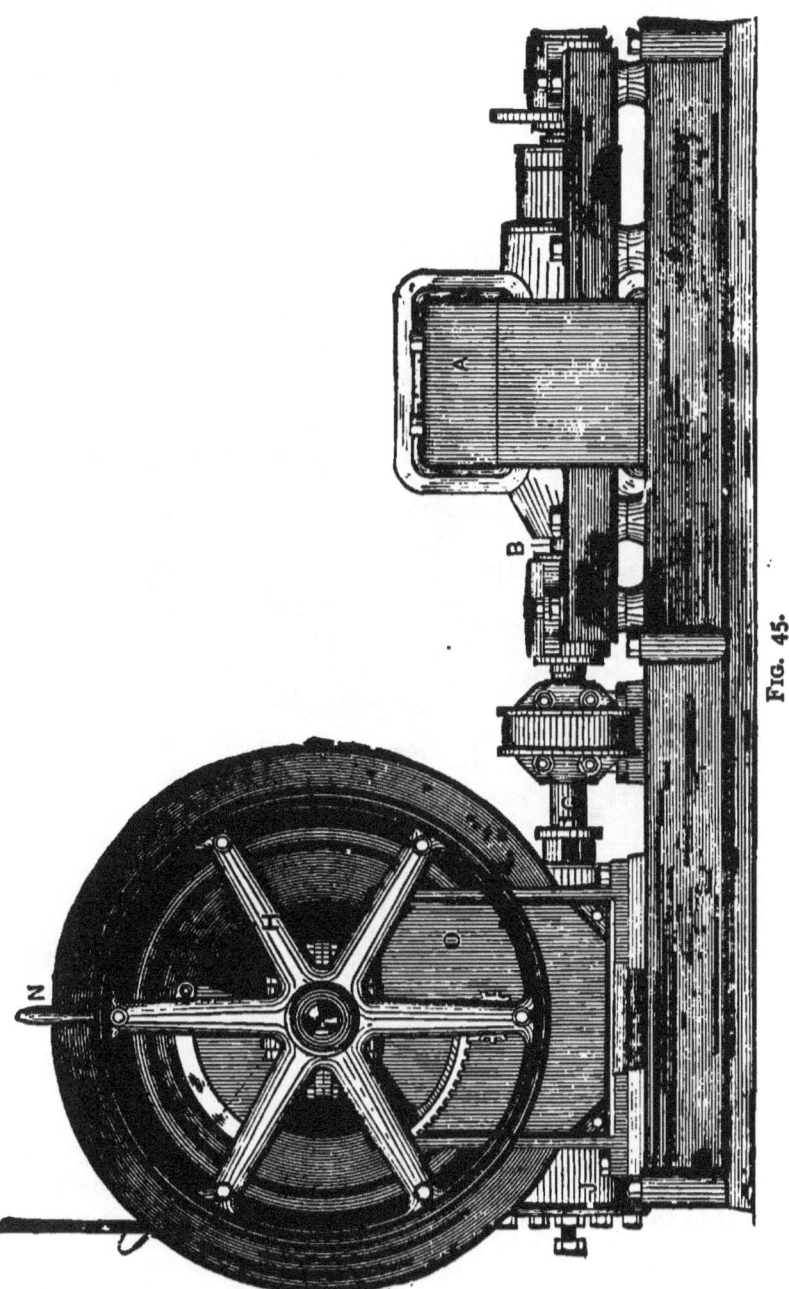

Fig. 45.

106 *THE ELECTRIC TRANSFORMATION OF POWER.*

which moves the switch over the contacts by the quadrant and pinion shown in Fig. 45. When the lever is vertical, the

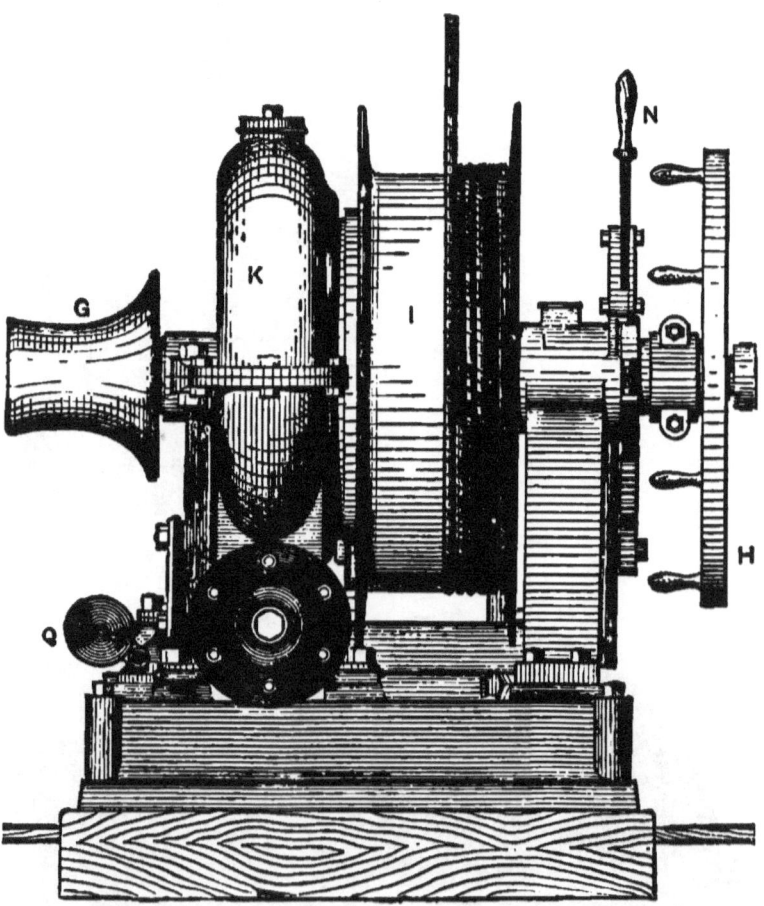

FIG. 46.

switch being in the neutral position, the current is turned off, but when the lever is in the position indicated by the dotted lines, the full current is flowing through the motor ;

APPLICATIONS OF THE STATIONARY MOTOR. 107

reversal of rotation not being required, reversal of current is unnecessary.

When freight is to be hauled and elevated, the hauling rope being given two or three turns round the winch-head and the current turned on, the winch-head, shaft, and hand

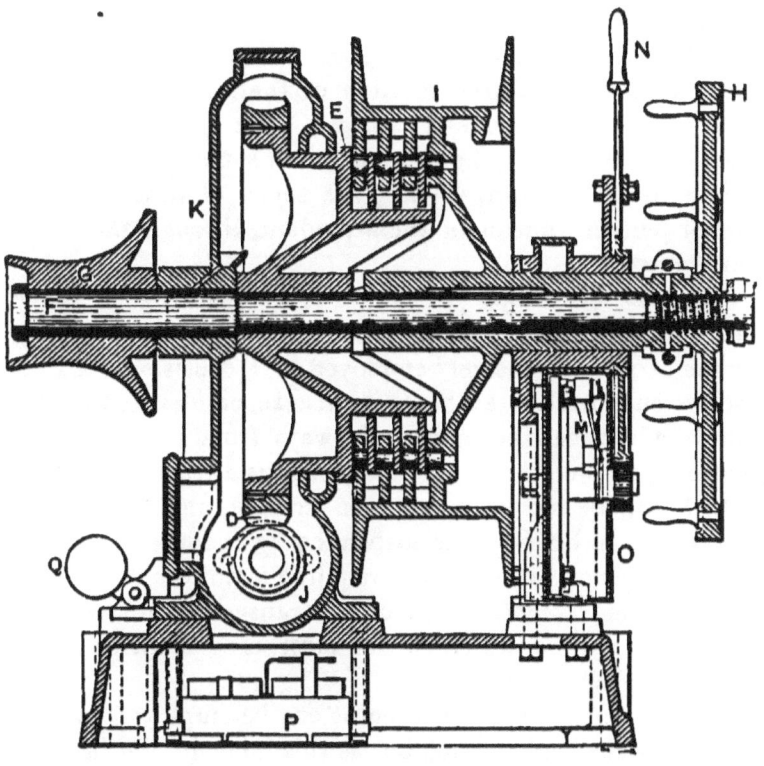

FIG. 47.

wheel H rotate, while the drum remains stationary: the load being thus hauled into position and the hoisting rope attached, the attendant stops the rotation of the wheel H by holding one of its handles; the rotation of the shaft continuing, screws the drum to the left against the friction clutch and puts it in rotation, the hand wheel also rotating and obliging the attendant to release his hold; and when the load has attained its elevation the current is turned off, and the rotation of the motor stops, the gearing on its shaft holding the drum shaft stationary; reversal of rotation by the load being prevented by the automatic brake Q, applied to the motor shaft.

When freight is to be lowered no current is required during its descent; the rotation of the hand wheel being reversed, so as to move the drum a little to the right and ease the pressure on the friction clutch sufficiently, the load descends by its own weight; the speed of descent being under perfect control of the attendant, through the wheel and clutch.

Electric Travelling Cranes.—For the moving of heavy loads in large foundries, mills, and factories, the traveling crane is now extensively employed. It consists of a steel bridge supported at each end on trucks, on elevated tracks, on which it can be moved sideways from one end of a building to the other. On this bridge is a truck, or trolley, which can be moved on a track across it, and from which the article to be moved is suspended by chains, by which it can be hoisted or lowered as required; and for these three kinds of movement, longitudinal, transverse, and vertical, power is required, which can be most conveniently applied by three electric motors.

The ordinary high speed motor can be employed and the speed reduced by gearing, but as this involves considerable loss of power, low-speed motors are preferable. The Brush Electric Company manufactures a motor for this purpose, the speed of which is only 200 revolutions per minute for 10 horse-power, and 100 for 5 horse-power, with an efficiency equal to that of a high-speed motor; and has equipped a traveling crane in its factory with three of these motors, which shows how they may be applied. A 10-horse motor is mounted on the truck at one end of the bridge, and its armature connected with a shaft which extends across the bridge, and carries, at each end, a pinion geared to a larger wheel connected with the truck wheels, by which the longitudinal movement is produced; and the other two motors, a 10-horse for the transverse movement, and a 5-horse for the vertical, are mounted on the trolley which traverses the bridge. The electric energy is supplied by a 500 volt dynamo.

Cranes of this kind are constructed having a maximum carrying power of 20 to 25 tons, and a span of 20 to 75 feet; and by mounting two on the same tracks, the total carrying power may be increased to 40 or 50 tons; so that the heaviest locomotives, castings, or ladles full of melted metal, can be moved with ease and safety to any required position.

The electric motor is applicable to any other kind of crane for which power is required.

Electric Operation of Printing Presses.—No special devices of any kind are required to adapt the electric motor to the operation of the printing press, except that double reduction of speed by countershafts and belting is usually necessary, but otherwise the application is just as simple as that of the steam engine or any other direct source of power. The rotation of the motor is in the same direction continuously, and a belt on the armature pulley, connecting it with the main shaft, through the countershafts, supplies the power to all the presses and other machines in a large establishment. The convenience and economy of this application of the motor are best illustrated by its practical operation as shown in a number of printing establishments in Chicago, visited by the writer in February 1892.

In one of these, electric power had been in use about two and a half years, supplied by a $7\frac{1}{2}$ h. p. Thomson-Houston motor, which was running 1 cylinder press, 3 jobbing presses, and a paper cutter; and was previously furnishing power also for running an equal number of presses in an adjoining office. The operation of the motor was perfectly satisfactory; running continuously ten hours a day, without noise, sparking, or failure; requiring no other attendance than a daily supply of oil, and filing of the copper-clad carbon brushes, not oftener than once a month.

The cost of the electric power as compared with that of direct steam power, supplied from an external source and paid for in a similar manner, was considered much greater; $22.25 a month being the average cost; while $12.00 a month was the estimated cost of steam power, which however was not available in this case; while the cost of generating the required steam power on the premises was estimated to be far in excess of the cost of the electric power supplied.

In another establishment a 10 h. p. Eddy motor was running 3 cylinder and 3 jobbing presses, besides a paper cutter and stitching machine, and giving perfect satisfaction both as to cost and quality of service. Its operation was noiseless, and free from sparking at the brushes, which were of bare carbon.

In a large establishment in which electric power had been in use for 18 months a 15 h. p. C. & C. motor was running 10 cylinder and 9 jobbing presses, besides extensive binding machinery. The motor service was highly satisfactory and its cost, which was about $134.00 a month, was estimated to be about the same per horse-power as that of steam. Copper brushes were in use, and the sparking, while not excessive, showed a marked contrast to its entire absence in the motors already mentioned, where carbon brushes were employed.

Power, in another large establishment, was supplied by an Edison 24 h. p. motor, at night; 17 cylinder presses and various other machines being thus operated at an average cost of $109.00 a month; which, it was estimated, was less than half what it would cost to do the same work by the steam power generated on the premises and employed for running the machinery and heating the building during the day. The motor had been in use four months, and its service was entirely satisfactory.

A 25 h. p. Beldin motor in another establishment was running 12 cylinder presses and 7 jobbing presses, besides various auxiliary machines; at a cost of $100.00 a month. Electric power had been employed for two and a half years, during which there had been a change of motors, and was satisfactory when in operation, but subject to occasional stoppage, the cause of which, as subsequently learned, was due to habitual carelessness, as was sufficiently evident from the fact, that the motor was found running with a partially consumed, red hot carbon brush.

The operation of the newspaper press by electric power is one of the most important applications of the electric motor. The Inter Ocean of Chicago is printed in this way, and the magnitude of the work performed will be understood when it is stated that its average daily issue, in May 1893, including the weekly and Sunday editions, was over 100,000 copies, averaging 16 pages each, and containing $2\frac{1}{4}$ square feet of printed matter to the page; and the entire work of printing these 1,600,000 pages, and cutting, folding, and counting out the 100,000 papers daily, was done automatically, without the touch of a hand; each press combining in itself the machinery for these various operations.

As the ground on which the Inter Ocean building stands is limited and very valuable, economy of space is of great importance; hence the employment of electric power was a matter of necessity; as the location of a steam engine and boiler of the required size, in addition to the massive presses, on the ground floor, where a solid foundation for the operation of such heavy machinery could be obtained, was found to be impracticable.

Electric power had been in use for nearly three years, and was then supplied by seven motors at a cost of about $188.00 a month; three 15 h. p. motors and one 25 h. p. running the four presses, and one 15 h. p. kept in reserve; one 7½ h. p. motor running the stereotype department, and another of the same size, the etching department; all, except the last, Mayo motors. Each of these motors is about 3 feet high and occupies about 9 square feet of floor space.

The service was satisfactory in the highest degree, not only in its cheapness and economy of space, but in its cleanliness and freedom from the many annoyances incident to the generation of steam power on the premises; so that even if such generation were practicable, the company say, that the electric service would be preferred, though its cost were doubled.

The assured success of the electric system, in this instance, has induced other newspapers to adopt it; and the Chicago Tribune has added electric power as an auxiliary to its steam power and a reserve in case of accident; employing it part of the time each week, running three of its presses with one 30 h. p. Mayo motor, and three others with three 15 h. p. of the same make, one to each.

Boyce's Weeklies, (the Saturday Blade, Chicago Ledger, and Chicago World,) eight-page papers, over 500,000 of which are issued weekly, are printed on three presses run by a C. & C. 30 h. p. motor, at a cost for current estimated at about $100.00 a month; the presses being employed ten hours a day. There is also a C. & C. 3 h. p. motor employed in the etching department and an Eddy 10 h. p. motor in the stereotyping department. Electric power, in July 1893, had been in use about two years, during which it had been largely increased and was giving satisfaction.

The Daily News has adopted electric power for the performance of various kinds of auxiliary work; its presses being run by steam power. Ten Eddy motors of various sizes, ranging from 3 to 20 horse-power, and representing, in the aggregate, 62 horse-power, are employed $16\frac{1}{2}$ hours a day, on an average, to run the etching and type-setting machines, drill press, lathe, dummy elevator and ventilating apparatus; the electric energy both for power, light, and photographic work, being generated on the premises by the steam power; an arrangement for supply of electric energy from a central station, in case of accident, being also provided. The motors, in Feb. 1892, had been in use about four months, and their service had been found eminently satisfactory.

Commercial Measurement of Electric Energy.—The electric energy in all these cases, including the provisional supply for the Daily News, was furnished from an Edison central station; the average distance to which it was transmitted being about a quarter of a mile; the electric pressure at the station being 228 volts, and at the various points of delivery, about 225 volts, allowing 3 volts for average fall of potential. A monthly test with an ammeter gives the maximum current in amperes delivered to the motor at the average 225 volt pressure, or E. M. F., and the electric energy being calculated at a special horse-power rate of 900 watts, equal to 4 amperes multiplied into the 225 volts, it is only necessary to divide the maximum current by 4 to obtain the maximum horse-power consumption; which gives the same result as multiplying the current by 225 and dividing the number of watts thus obtained by 900.

The consumer is entitled to the use of the maximum current ten hours a day, at a stipulated monthly rate for different classes of work, as continuous, intermittent, and elevator work, estimated according to the horse-power of his motor; and an Edison current meter shows, by the quantity of zinc chemically deposited, the actual daily current consumption, and hence any excess in the use of current above this stipulated number of hours.

Electric Operation of Dental Apparatus.—The electric motor furnishes the most convenient and economical means of operating the various mechanical appliances of a dentist's office, where current for this purpose is available; an $\frac{1}{8}$ or $\frac{1}{12}$ h. p. motor supplying sufficient power for the requirements of an ordinary office, at a cost of about $3.00 a month, when current is supplied from a central station.

The principal appliances thus operated are the dental lathe, for plate work in the laboratory, and the dental engine, in the operating room, which runs the drills employed in preparing cavities for filling, and the brushes and sand paper used for cleaning, scouring, and polishing the teeth. The latter instrument is constructed with a light, flexible, cable, made of steel wire spirally wound, by which rotary force may be applied. To this is attached a small hand chuck, in which the drills and other apparatus are inserted, and which may be held at any required angle. These engines are of two kinds, known respectively as the foot, and the suspension engine; the former operated by a treadle, attached to a base from which projects a vertical rod, which supports the drilling apparatus; and the latter attached to the ceiling, so as to suspend the drilling apparatus in proximity to the operating chair, and connected with any available source of power.

No special device is required for the electric application of power to the lathe, which is run by an ordinary belt connection with the motor pulley; but the engine, from the peculiarly delicate nature of its work, must be under the most perfect control of the operator, be capable of being run at any required rate of speed, fast or slow, in either direction, stopped instantaneously, or its rotation reversed;

and hence requires special adaptation to these various conditions.

All this is accomplished in a very satisfactory manner by a method recently devised by the Edison Company and now in successful operation. The field-magnet of one of their small, slow speed motors is wound externally with three or four layers of iron wire, which is connected in series with the copper coil, and tapped at five different points, so as to divide it into five sections of equal resistance, connected in series; and these are connected by a flexible cable with nine contact stops, arranged in an arc on a small insulating switch-board, in such a manner that when the switch, which forms part of the field circuit, is on the central stop, all this resistance is cut out, and maximum current, and hence maximum speed obtained; but when the switch is moved from stop to stop, in either direction, the resistance coils, one after another, are included in the circuit, and the speed proportionally reduced; and when the switch rests on a blank stop, provided at each extremity of the arc, all the coils are cut out and the rotation ceases.

On the reverse side of the switch-board are four contact stops, two near each edge, about half an inch apart, connected through the cable with opposite poles of the armature circuit; a pole-changer, consisting of a short brass bar which forms part of this circuit, makes contact by spring pressure with two of these stops at a time, one at each end; so that when moved to the two adjacent stops, the armature current, and hence the rotation is reversed. Two short insulating spurs, attached to opposite ends of this pole-changer, project through slots in the switch-board, so as to make contact with the switch, in front, at each end of the arc; so that if the switch is moved far enough at either end of the arc, to push one of these spurs to the end of the slot, the pole-changer is shifted and the rotation reversed.

When slow stoppage without reversal is desired, the

switch is moved to the blank stop, at either end; if reversal is required, it is moved half an inch farther, shifting the pole-changer; and if instantaneous stoppage is required, it is first moved so as to shift the pole-changer, and then back to the maximum contact stop, so as to produce a momentary, full, reverse current, and then to a blank stop; all of which can be done in an instant by a quick double movement.

The switch-board may be connected with the treadle of the foot engine, and the switch movement controlled by foot pressure, or mounted in any convenient position for hand control, and connected with the suspension engine.

The motor may also be employed in connection with the dental plugging instrument, used for compressing gold fillings. This instrument is operated by condensed air supplied through a rubber tube by a small condensing pump, run by a crank which produces the reciprocating motion of the piston in the usual manner; so that a belt connection with the motor is all that is required.

Electric Operation of Medical and Surgical Apparatus.— Medical treatment by static electricity is given by plate induction machines of high E. M. F., requiring rotary motion of moderate speed, which are most conveniently operated by light power, which may be furnished by a small electric motor where current is obtainable; the motor being either attached to the bed of the machine or placed in any other position convenient for belt connection. An instrument for treating the throat by the injection of a medicated spray, by pneumatic pressure, requires connection with a chamber in which air is condensed by a small pump, which may be conveniently operated by an electric motor; a $\frac{1}{18}$ horse motor furnishing sufficient power both for the static machine and this apparatus.

The extensive heavy apparatus of various kinds employed in giving massage treatment by the aid of mechanical power, may be operated by a motor of sufficient horse-power for the requirements of the establishment, without the intervention of any special electric device; a constant rotary force in the same direction, applied to a shaft, being all that is required; which, by mechanical devices at the various machines connected with this shaft by belts, is made to produce the vibratory or other movement required by the treatment; the resistance by which speed is controlled being furnished by contact with the patients receiving treatment.

The motor is also applied to the operation of various surgical instruments, employed in drilling bone, and in sawing for the removal of dead bone. For the latter purpose a circular saw is operated by a small, globular shaped motor, of special construction, to which it is attached; which can be held in the hand so as to apply the saw to the limb or other part; current being supplied to the motor by a flexible conductor.

Electric Operation of Ship Drills.—In the construction of steel ships a large amount of drilling is required which can be most conveniently performed by the aid of an electric motor. For this purpose a motor of sufficient power to run a rotary drill is mounted on a truck by which it can be moved to any position about a ship on the stocks where drilling is required, and connected by a flexible conductor with a generator at the factory.

By gearing mounted on the motor, its armature is geared for reduction of speed to a Stowe flexible shaft about 11 feet long, which, at its outer extremity, is geared to the drill press by which the rotary force is applied; the flexible shaft permitting such change of position as may be required for drilling several adjacent holes. A switch mounted close to the drill press, and connected with a rheostat enables the operative to control the current.

The Edison Electric Percussion Drill.—For drilling rock or mineral, a drill having a reciprocating movement by which blows may be struck in rapid succession, in making a boring, is required. It is known as a percussion drill, and in mines, where the employment of direct steam power is prohibitive, it has heretofore been operated by pneumatic pressure, and has been one of the chief obstacles to the introduction of electric power in mining, since the pneumatic system was necessary for the operation of this drill, so long as there was no similar electrically operated drill which could take its place; and hence the employment of pneumatic power exclusively was found to be the most economical. The invention of an electric percussion drill combining qualities which render it superior to any similar instrument operated by pneumatic pressure or otherwise, has removed this obstacle, and made the exclusive employment of the electric system in mining, for power as well as for lighting, the most economical and most convenient.

The Edison drill for this purpose is constructed, as shown in Figs. 48 and 49, with an iron cylinder 38 inches long and 7 inches in diameter, which contains two solenoid coils of insulated copper wire of medium gauge, both wound in the same direction, each about $8\frac{1}{2}$ inches long, $6\frac{1}{4}$ inches in external diameter, and $2\frac{1}{8}$ inches in internal diameter, which inclose a wrought iron plunger, about 14 inches long and 2 inches in diameter, so mounted as to have a free reciprocating motion without contact with the coils. To each end of this plunger is rigidly attached an aluminum bronze bar of the same diameter; the front bar, which is about 13 inches long, projects from the cylinder through an elongated bearing and carries, at its outer extremity, a chuck to which the bits are attached; and the rear bar is milled spirally for a distance of about 9 inches, and moves through a hexagonal bearing made to fit this spiral in the center of a ratchet wheel, by which a twist of one sixth of a turn or less is given to the bar at each stroke, varying according to its length; the ratchet wheel preventing opposite rotation as the bar returns, so that the bit cuts at a fresh angle at each stroke; the length of stroke being regulated by the hand screw shown at the side by which the drill is fed forward on a slide as the boring deepens. By inverting the ratchet

FIG. 48.

wheel and thus rendering the rotary apparatus inoperative, the drill may be employed to operate a blunt instrument for breaking out holes in rock. In the rear end of the

120 *THE ELECTRIC TRANSFORMATION OF POWER.*

cylinder is a spiral steel spring, against which a steel buffer, carried at the outer end of the rear bar, strikes; the mechanical energy thus absorbed from the back stroke being imparted, by the recoil of the spring, to the forward stroke, whose force is thus increased.

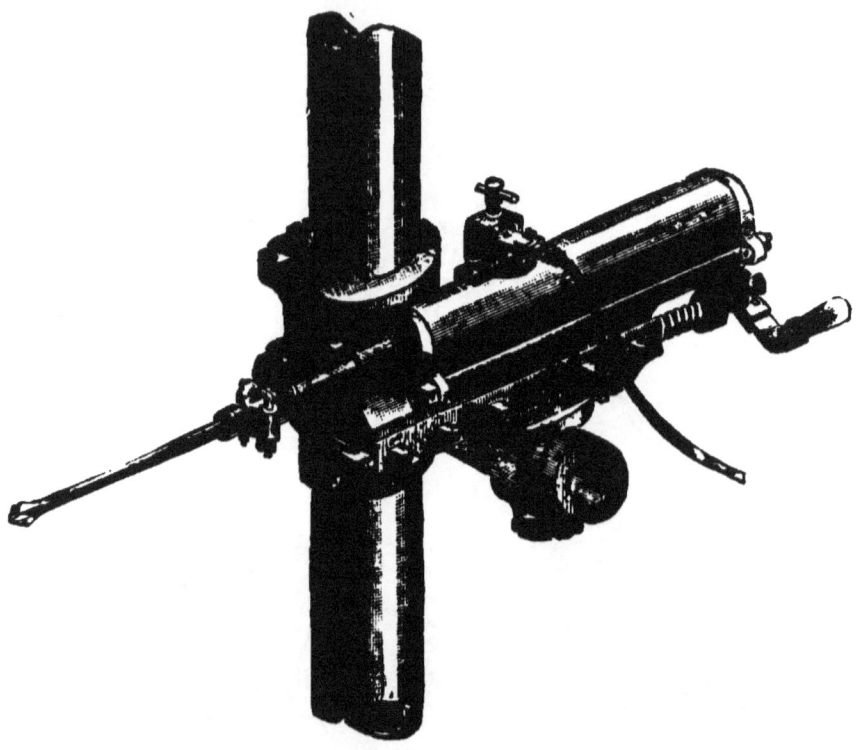

FIG. 49.

The reciprocal movement is produced by the alternate opposite attraction of the two solenoid coils which are placed end to end near the front part of the cylinder, and whose action is reenforced by the coefficient of magnetic induction in the inclosing iron cylinder, which makes them electromagnets. Just above their junction are three contact plugs, mounted on top of the cylinder outside, in a socket having a set-screw, as shown. The inner terminal of each coil is arranged for contact with the central plug, and the outer for contact with a separate one of the end plugs. A flexible cable, connected with a dynamo, contains three insulated copper conductors, stranded for flexibility, which terminate in a connector, in which they are arranged to make proper contact with these three plugs, when it is thrust into the socket; and when this connector is screwed down against spring pressure below, the plugs are brought into contact with the coil terminals, and the circuit being thus closed, current is transmitted alternately, in opposite directions, to each of the coils through the end plugs, returning by the central plug. The iron bar to which the drills are attached, acting as an armature, is therefore alternately attracted in opposite directions in this magnetic field. The direction in which the current flows, whether positively or negatively, and in which the coils are wound, whether similarly or oppositely, is of no special importance, since the alternate attraction of the bar is not dependent on either of these conditions.

To produce this alternating current, a series wound dynamo of the ordinary direct current type, separately excited by an auxiliary dynamo, is furnished with two collectors, insulated from each other, instead of the commutator; one a complete ring on which a single brush, connecting with the central conductor makes contact, and the other a half ring on which two opposite brushes, connected with the other two conductors, make contact alternately; one brush being insulated while the other is in contact. These collectors being connected respectively with opposite ends of the armature circuit, positive current enters the half ring from one of the brushes and negative from the other, alternately at each revolution of the armature, traversing each solenoid coil of the drill alternately, and returning by the complete ring.

This drill is therefore, practically, an electric motor of the simplest construction, operated by an alternating current, and having a reciprocating, instead of a rotary motion, and generates counter E. M. F., like other series motors, which varies inversely as the speed. Hence if it is allowed to run without load, that is without drilling, the speed and resulting counter E. M. F. are greatly increased, and the direct E. M. F. proportionally neutralized, reducing the stroke to a mere vibratory motion of about an inch in range; but when applied to work, these conditions are reversed, and the force of the stroke increased by the relative increase of direct E. M. F. If therefore the drill is not fed up sufficiently for the bit to strike, it immediately assumes this vibratory motion, and may be allowed to run in this manner as long as desired without injury, since it merely vibrates in space without striking anything; the length of its range being circumscribed by the alternate, opposite, electromagnetic force, which acts as a cushion on the plunger. *The drill may also be moved so close to the boring as to allow no range of stroke, and left so, as long as required, without injury, with the current flowing through the coils.

The plunger has a range of from 3 to $4\frac{1}{2}$ inches, and makes about 600 strokes a minute; and the stroke can be shortened to $\frac{1}{4}$ of an inch, if required in starting a hole. At the normal stroke, with a $1\frac{1}{2}$ inch bit, the boring can advance at the rate of 2 inches a minute in the hardest granite, consuming about 3 horse-power; the feed being arranged for a depth of about 20 inches.

The drill weighs about 400 pounds; its two copper coils being about 60 pounds each. It may be mounted on a tripod, as shown in Fig. 48, the three weights of which are about 100 pounds each, or attached to a column, as shown in Fig. 49, or shifted from one to the other as required; and can be operated at any required angle, or in any available position.

Several drills can be run on parallel circuits derived from a main circuit connected with a single dynamo; the stroke of each being in synchronism with the alternations of current at the dynamo, and hence occurring in all simultaneously. The circuit of any of these drills may be opened by reversing the set screw, and its supply of current cut off without detriment to the other drills.

No electric knowledge or skill is required to operate this drill. It can be started by simply thrusting the connector into its socket and screwing it down so as to close the circuit, without any liability of wrong connection; and can be taken apart and put together again, in running order, in a few minutes by any ordinary workman, or miner. It has no delicate electric parts or connections to get out of order; no packed joints or tight fittings; requires no special precautions against dripping water, since its electric coils are securely inclosed; nor any special insulation to prevent electric waste, or protection against dangerous currents, since its E. M. F. is so low as to insure freedom from electric leakage, and perfect safety to the operatives. There is no sparking, and hence no danger of igniting inflammable gas; and no danger of burning out. It will bear rougher usage than any drill operated by other power, and its few wearing parts are easily and cheaply replaced when worn out.

The Van Depoele Electric Percussion Drill.—This drill, manufactured by the Thomson-Houston Electric Company, is similar in external appearance and general construction to the one last described, but differs materially in its electric construction and operation. The cylinder incloses three copper coils, placed end to end, as shown in Fig. 50; the central coil, Y, termed the polarizer, composed of

124 *THE ELECTRIC TRANSFORMATION OF POWER.*

numerous convolutions of small wire, which therefore carries a relatively small current, is employed chiefly to magnetize the iron-plunger and keep it magnetically saturated; while the two end coils, X and Z, composed of comparatively few convolutions of large wire, which carry a large current, are employed to operate the plunger.

The electric connections of these coils with each other and with the dynamo, E, located at the power station, are

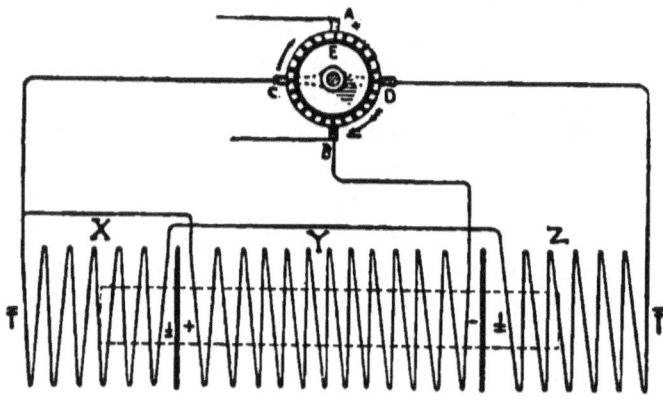

FIG. 50.

as follows:—Two pairs of brushes, A and B and C and D, make contact on the commutator, A and B being stationary, while C and D are made to rotate in the same direction as the commutator, over its surface, but at much slower speed, by means of a counter-shaft, belted to a circular brush-yoke. The coils X and Z, which are connected together in series, as shown, are connected by their terminals to the two rotating brushes C and D, while the central coil, Y, has one terminal connected to the negative stationary brush B, and the other, which thus becomes the positive, to the rotating brush C, as shown. The stationary brushes are placed, as usual, on the neutral line, indicated here by the vertical position, and therefore collect the current at its highest potential, 220 volts, while the current collected by the rotating brushes varies from zero, when they are in the horizontal position, at right angles to the neutral line, to the maximum, 220 volts, when they are in the vertical position, and is alternately reversed twice during each revolution, as their relative positions with reference to the neutral line are reversed. This is therefore a pulsating current, rising and falling in waves, which flows alternately in opposite directions through the coils X and Z.

The current through the central coil, Y, flows continuously in the same direction from right to left, as indicated by the signs minus and plus; rising to its maximum potential of 220 volts, when the rotating brush C, with which its positive terminal is connected, is on the neutral line, either above or below, and falling to its minimum of 110 volts, when this brush is on the horizontal line, since its E. M. F. is then only equal to the 110 volts potential difference between the stationary brush and zero of the rotating brush, instead of between 110 positive and 110 negative, as when the latter brush is on the neutral line. It is therefore a pulsating current like the other, ranging from 110 volts minimum to 220 maximum; and since it is constant in direction, the polarity of the plunger, which is magnetized by it, remains unreversed, and is in accordance with the direction of the winding, which we may, for convenience, assume to be such as to produce a north pole at the right end and a south pole at the left.

When the rotating brushes, C and D, are in the position shown, the E. M. F. in the end coils being neutralized, the current, which is then at its minimum E. M. F., is confined to the central coil, the plunger is therefore drawn into this coil, as shown by the dotted lines. As C and D rotate toward the vertical position, current flows through the end coils, and its E. M. F. increases in all the coils, attaining its maximum when these brushes are on the neutral line. All the coils have then attained maximum polarity, which in the coils X and Z is the same at the corresponding ends of each. But as these two coils are in series with each other, north polarity predominates in one and south polarity in the other, north polarity increasing in the direction of current flow and south polarity in the opposite direction, so that these two coils must be regarded as practically one coil, having north polarity at one end and south at the other. Each pole of the magnetized plunger is therefore attracted by the nearest opposite pole of the magnetized coils and repelled by the nearest similar pole; the plunger being thus moved into the position in which it can embrace the greatest number of lines of magnetic force. Reversal of the current through these two coils reverses their polarity, moving the plunger, whose polarity remains unchanged, in the opposite direction, and thus producing the reciprocating stroke of the drill.

As the current in the central coil flows from right to left, its magnetic energy is added to that of the end coils when the stroke of the plunger is in that direction, and opposes it when the stroke is in the opposite direction; hence the force of the right to left stroke exceeds that of the left to right stroke by double the magnetic energy of the central coil.

In drilling hard rock it is important that the principal force be concentrated on the forward stroke, the back stroke requiring only sufficient force to move the plunger; but in drilling softer material the bit is liable to stick in the deep cut made at each stroke, requiring greater force to draw it back than to urge it forward; in which case it is important to concentrate the principal force on the back stroke. Fig. 51 shows how the principal force may be made reversible for either purpose, by reversing the connections of the central coil; its left hand terminal being

APPLICATIONS OF THE STATIONARY MOTOR. 127

connected with a switch by which connection may be made with either of the rotating brushes, the connection of the right hand terminal being, at the same time, changed to the opposite stationary brush, as shown; the stationary brushes being, for convenience in making the drawing, placed on a horizontal line, which here represents the neutral.

By varying the speed of the rotating brushes and thus varying the frequency of alternation of current, the rapidity

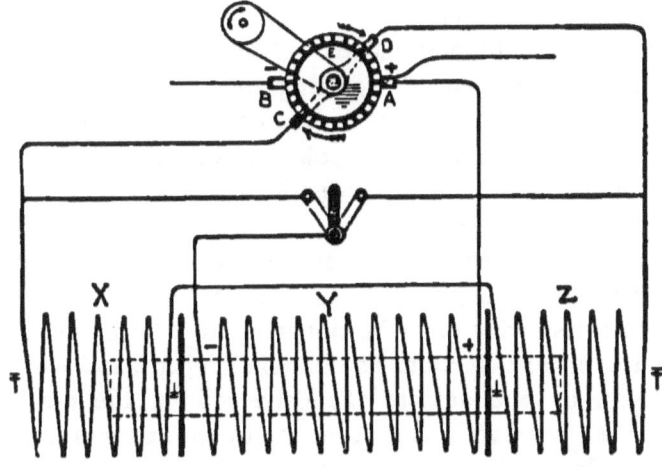

FIG. 51.

of the stroke can be varied to any extent required; ranging from 300 to 500 strokes a minute.

It is evident that the electromagnetic energy of a three coil drill operated in this manner, both by attraction and repulsion, acting on a polarized plunger with a force alternately varying in intensity, must exceed that of a two coil drill operated by attraction alone, acting on an unpolarized plunger with an alternating force of invariable magnetic intensity. It has also the advantage of reversibility in force of stroke, as shown. The two coil drill has, on the other hand, the advantage of greater simplicity and cheapness of construction.

This machine may be so modified as to be applied to the operation of any apparatus requiring a reciprocating force, as a pile-driver, power-hammer, or reciprocating pump. A light pump operated in this manner and occupying a limited space is employed in sinking shafts, and can be suspended and lowered as the depth increases. No close fitting piston and valves such as are employed in a similar sieam pump are required.

The Electric Diamond Drill.—This is a rotary drill employed in mining for prospecting, by which holes of any required depth up to 500 feet may be drilled and the core extracted for examination in solid pieces, showing the character of the rock or mineral, depth of mineral deposit, and stratification.

It is constructed with a steel core-barrel of convenient length for the reception of the core, attached to a hollow steel drill-rod, which may be elongated as required by screwing on additional sections as the hole deepens, and which is attached, by a chuck, to the drive-rod on the machine, while the core-barrel carries, at its outer extremity, a hollow circular steel bit, in the rim of which are set 6 or 8 black diamonds, by which a circular cut is made around the core, as the bit rotates. The holes vary in diameter from $1\frac{1}{2}$ to $2\frac{3}{4}$ inches, and the corresponding cores from $\frac{1}{16}$ of an inch to $1\frac{7}{8}$ inches. The core passes into the barrel, and when sufficiently long, is gripped by a split collar which slips down into a beveled chamber, as the bit is withdrawn, and breaks it off; the pieces varying in length from a few inches to several feet.

The drill complete is shown in operation in Fig. 52; the drive-rod with drill-shaft attached being shown at the right above, with its central gear by which the rotary force is applied, and differential gearing by which the friction feed is operated. To the rear end of this drive-rod is attached a hose connected with a force-pump on the left, by which water is constantly injected into the hole through the tubular drill-rod, and the loose borings thus forced to the surface. At the right corner below, is shown a hoisting drum with a coil of wire rope, employed to pull out the drill-rod with attached core-barrel and bit. In the center

FIG. 52.

is shown the electric motor by which all this apparatus is operated, the speed being reduced by gearing to about 600 revolutions of the drive-shaft per minute; the rotary motion when applied to the pump being made reciprocating by a crank. Above the motor is shown the switch-board, on which the electric connections are made and the cut-out switch mounted; under which is the rheostat, and on the right, the lever which moves the contact switch by which the circuit through the rheostat is closed and opened.

In practical work done with this machine, with 3 horse motor power, by the Aspen Mining Co. in Colorado, in hard limestone, intersected with shattered strata, bands of tough siliceous rock, and clay seams, holes of $1\frac{1}{2}$ and $2\frac{3}{4}$ inches diameter were drilled at an hourly rate of 9 to 42 inches; the average for constant work in a single month, Feb. 1891, being 1.66 feet per hour, at an exact cost, including all expenses, of 68 cents per foot; the work being done underground, in old drifts, levels, and slopes, in which much time was consumed in moving and setting up the drill. This was the first work done with an electric drill of this kind, and its cost has since been considerably reduced.

The Triplex Electric Pump.—The pump is one of the most important pieces of apparatus required in mining, as the presence of water is one of the principal difficulties met with, and efficient apparatus for its removal is an absolute necessity. As such apparatus must be adapted to limited underground space, and often requires motive power of high efficiency, the electric motor, combined in the same machine with the pump, has special adaptation to the required conditions.

The Gould triplex pump combined with the Thomson-Houston electric motor, as illustrated in Fig. 53, is constructed on these principles. It has three vertical cylinders, each inclosing a piston attached in the usual manner to a crank on the main shaft above. The three cranks being arranged at relative angles of 120°, produce uniform load on the motor under all conditions, without dead centers.

Double reduction of motor speed is required, which is obtained by two sets of gears, as shown; the armature shaft being geared by a small pinion to a large gear wheel on the left, which is mounted on a shaft similarly geared, on the right, to the crank shaft of the pump, by which its speed is reduced to a range of 35 to 50 revolutions a minute.

APPLICATIONS OF THE STATIONARY MOTOR. 131

These pumps range in capacity from 25 to 250 gallons of water per minute, pumped from depths of 100 to 300 feet, employing motors of 1 to 15 electric horse-power.

For raising water from greater depths, a pump of stronger construction is required, for which the horizontal form is usually preferred, and in which the pistons are attached

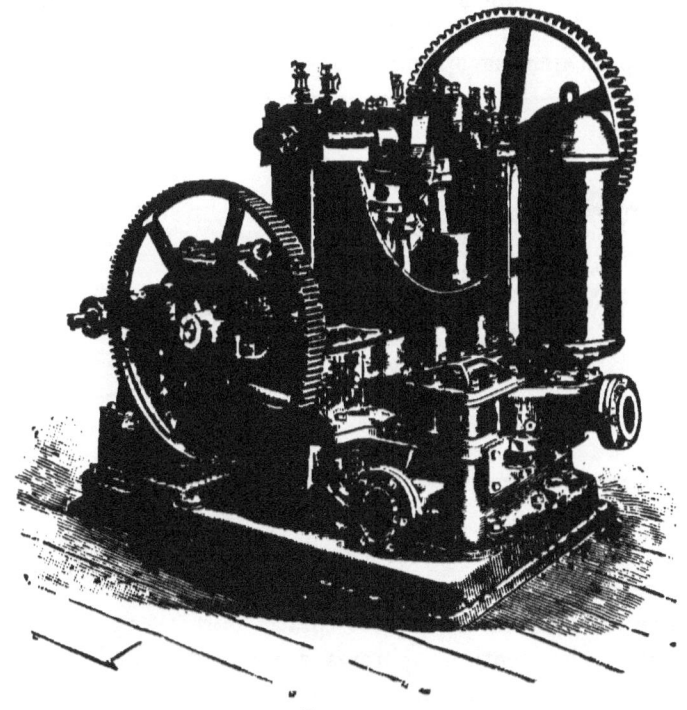

FIG. 53.

to a cross-head, moving in guides, by which they are kept in alignment with the cylinders. It may be either triplex or duplex, the strain on the motor being approximately steady with either kind; the winding of the field-coils being varied somewhat for adaptation to the special conditions of service in each particular case, and the horse-power of the motor made proportional to the load, or head of water. With a Knowles high pressure duplex pump combined with a Thomson-Houston 60 horse-power motor, now in successful operation in a Michigan mine, water is pumped from a depth of 1500 feet, at the rate of 100 gallons a minute.

The Sperry Pick Electric Coal Cutter.—This machine, illustrated by Fig. 54, is constructed with a reciprocating shaft carrying a pick at its outer extremity, and surrounded by a spiral driving spring inclosed in a cylinder, which is

FIG. 54.

mounted on a two wheel truck, which also carries the motor and gearing.

The armature of the motor rotates continuously, and, by means of the gearing, operates a retractor consisting of a wheel, with a projecting pin which engages a catch on the rear end of the shaft, at each rotation, pulling the shaft back from $6\frac{1}{4}$ to $7\frac{1}{2}$ inches against the force of the spring, and then releasing it to be projected forward by the recoil of the spring, which drives the pick into the coal.

The spring is made in sections and requires a retractile force of about 125 pounds to the inch for its compression, the initial force being about 500 pounds, and the total compressing force about 1375 pounds.

The armature gearing is provided with a spring cushion, and a similar cushion is introduced between the retractor and the main gear to relieve the concussion and strain on the motor and gearing produced by the percussive blows of the pick. These blows are delivered at the rate of from 160 to 225 per minute, with a mean effective force of about 1000 pounds.

The weight of the shaft varies from 100 to 150 pounds, according to the size of the machine, and requires for its operation a motor of about four horse-power, with a current of about ten amperes and an E. M. F. of 220 volts. The machine is furnished with handles, at the rear, and can be moved and operated by one man.

The New Arc Electric Coal Cutter.—This machine is designed for making a cut under the coal, which then usually falls by its own weight or may be removed by blasting, when necessary. It is constructed as shown in Fig. 55, with a heavy iron base, cast in one piece, which serves as a foundation for all the parts. On this are mounted the motor, shown near the center on the right, and the shafts and gearing connected with it, by which the various parts are operated; also the cutter, feeding apparatus, and wheels by which the machine is moved along the track.

The motor is of 15 horse-power and supplies the power for moving the machine and operating its various parts, about 7 horse-power being employed in operating the cutter. In the latest machines it is made water-tight and thoroughly protected from external injury. The required E. M. F. being only 220 volts, no risk is incurred by the employment of the electric current required for its operation.

The armature shaft carries a beveled gear at each end, the one at the rear end being made to engage, by means of a lever, either one of two beveled gears connected with the transverse shaft shown on the right, by which it can be

134 *THE ELECTRIC TRANSFORMATION OF POWER.*

Fig. 55.

rotated in either one of two opposite directions. On this shaft is shown a worm gear which engages a gear whee underneath it, mounted on a shaft extending at right angles to the left, and there connected with a train of gearing which operates the feed chain. This chain passes over two friction pulleys, as shown, and is attached to a post in front; and by this means the machine can be moved at any degree of speed required for the cutting, in proportion to the hardness of the coal, or held stationary when necessary in cutting specially hard formations.

The same gearing, by means of another controlling lever, rotates a shaft, on the front end of which is shown a worm gear, which engages a semicircular gear by which the arm carrying the cutter is moved round from its first position, alongside the machine, at the beginning of the cut, to the position shown, and there held rigidly during the remainder of the cut.

The cutter consists of a series of knives attached to an endless chain supported on the arm, and passing round a drum at each end of it, which is rotated by gearing connected with the beveled gear on the front end of the armature shaft, as shown. The knives are shaped to cut alternately at different angles, as shown, and also rake out the coal, leaving a clean under-cut about three inches wide and from three to six feet deep; the depth being usually made the same as the thickness of the vein.

The average cutting rate is from 45 to 60 feet per hour, including track laying and adjustment of the machine; varying according to the hardness of the coal.

The length of the arm is adjustable within required limits by means of the block shown at its outer end, in order to take up the slack produced in the chain by wear. The knives are easily detached from the chain for renewal, when worn out; and both knives and chain are drop forged and capable of withstanding the hardest usage.

The track along the face of the work consists of two rails connected by an iron cross-tie, and is easily removed and relaid. The machine has two sets of wheels, mounted on axles placed at right angles to each other. One set are flanged and adapted to the gauge of the permanent track of the mine, on which the machine can be moved to any required point; and the other pair, by which the machine is mounted on the movable track, are not flanged, but are held on the track by single and double guides, as shown in front of the wheels, two pairs of each. The support of the machine is easily changed from one set of wheels to the other by means of a cam operated by screws; the flanged wheels being raised out of the way as shown, when the machine is in operation.

This cutter is employed for what is known as "wall mining," where the coal is cut from a wall surrounding an area of convenient size; whereas the pick cutter is employed for "room mining," where the cutting is done in narrow rooms, between which are left supporting portions of coal.

Various Electric Mining Apparatus.—It is unnecessary to describe in detail the applications of the electric motor to the various other kinds of mining machinery, prominent among which are rock breakers, pulverizers, stamp mills, amalgamators, excavators, ramming machines, hoists, forges, ventilators, oil drills, and oil pumps; all of which can be successfully and economically operated in this manner. Electric tramways, tram-cars, and underground haulage will be described in the next chapter, in connection with electric railways and motors.

It is evident that when electricity has been introduced into the mine for the operation of machinery, it can be used for lighting also, and the superior advantages of the incandescent light for mining purposes, which have been fully demonstrated, obtained at a comparatively small additional expense.

CHAPTER V.

ELECTRIC RAILWAYS AND RAILWAY MOTORS.

General Remarks.—Traction by the electric transmission of power is eminently adapted to the operation of street railways, where its superiority to animal traction and cable traction in speed and economy, and to the latter also in facility and cheapness of construction, and in freedom from the frequent annoying interruptions incident to broken cables, has been amply demonstrated by the numerous electric roads now in successful operation.

The various details of the electric railway system have been made the subject of careful experiment for the last ten years or more, and its practical operation during the last five years has resulted in many changes and important improvements in the earlier methods, more particularly, of late, in the construction of motors having special adaptation to this service; so that different roads show important differences in the methods by which they are operated.

Line Construction.—The principal requisites of the system, as now generally operated, are the generation of electric energy by steam power at a central station and its transmission by conductors to electric motors on the cars. There are two principal methods of transmission on surface roads, known respectively as the conduit and the overhead methods. The latter is by far the cheapest and most practical, and hence the one generally adopted, while the former is demanded in cities where overhead conductors are prohibited, and has been made the subject of many costly experiments, but thus far with very limited practical success;

138 THE ELECTRIC TRANSFORMATION OF POWER.

though its promoters still hope that the obstacles may be overcome. These are chiefly due to the difficulty of obtaining proper insulation for bare wires in an open street conduit, liable to accumulations of mud, and to flooding in low ground, and also of maintaining contact with these wires.

The overhead method is illustrated by Fig. 56, which shows a single, bare copper wire, C of large gauge, properly insulated above the track JJ, at a suitable hight for the passage of the cars beneath it, and connected with a dynamo at the central station, G, by which a positive current is transmitted to the motors, MM, mounted on the trucks under the cars; the electric circuit being indicated by the dotted lines and the direction of the current by the

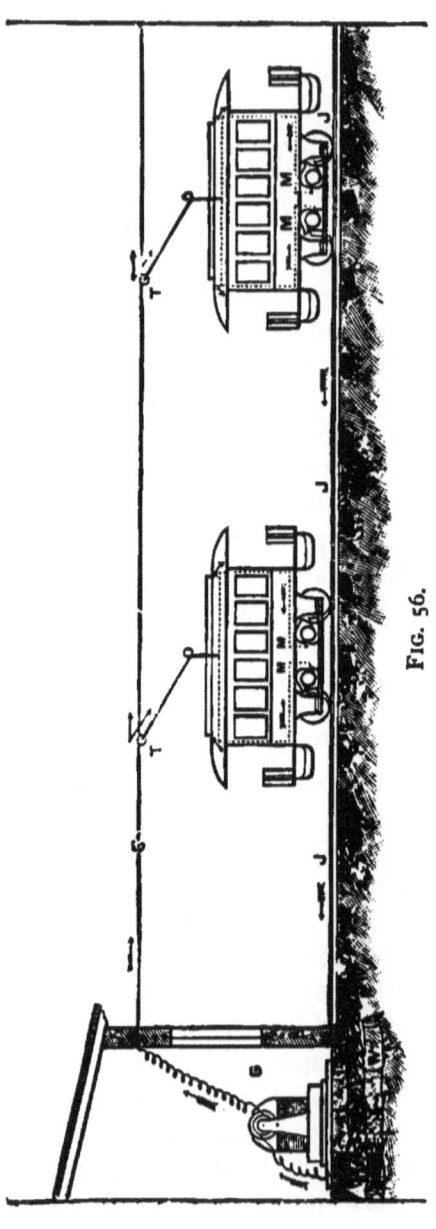

FIG. 56.

arrows; the negative current returning to the opposite brush of the dynamo by the car axle, wheels, rails, and copper conductor *W*. Proper electric connection between successive rails is made by bonds of copper or iron wire, shown at the junctions underneath, which are riveted or otherwise attached to the ends of the rails; similar connections being also made between opposite sides of the track at suitable intervals, so that both sides shall act as a single conductor. Iron wire is often preferred to copper for this purpose, to prevent the oxidation and consequent resistance to which junctions of two metals are liable from electro-chemical action.

The overhead conductor carries sufficient current to run all the cars on the line, to which it flows in parallel; each car taking so much as its motors require, and becoming the outer terminal of that section of the line between itself and the station, at any point where it happens to be: so that cutting off the current from the motors on any car does not at all interfere with the supply of current to the other cars; the total resistance of the circuit varying inversely as the number of motors taking current in parallel, becomes self-regulating, as in similar conditions in electric lighting; so that the strength of the current flowing to a single car, taking current alone, would, practically, be nearly the same as when flowing simultaneously to all the cars on the line.

As the transmission is dependent on the difference of potential, or E. M. F., between the overhead wire and the rails, it is quite immaterial whether the upper current is positive and the lower negative or the reverse; the liability to leakage being no greater in the one case than in the other.

Since the electric resistance varies directly as the length and inversely as the cross-section of the conductor, the power station requires to be located as near the center of distribution as possible, so as to reduce the resistance to the minimum and the size of the wire to the smallest gauge consistent with proper conductivity; No. o to 4 B. & S. gauge being the sizes most commonly employed.

Feeders.—On long lines, or those on which the grades are steep and the traffic heavy, feeder wires, wrapped for insulation, extend from the station and are connected with the line wire wherever a supply of current is required to compensate for the fall of potential due to the absorption of electric energy by the motors, and thus maintain the required normal E. M. F. on all parts of the line; the fall of potential in the feeders being only that due to the electric resistance of the wire. A feeder may carry current in parallel with the line wire and be connected with it at several different points; or it may carry the entire current and supply it to sections of the line wire insulated from each other; or several such feeders may extend to different sections, which may or may not be insulated from each other; two or more feeders sometimes supplying one section. The line thus acquires the requisite current carrying capacity, without abnormal size of the line wire.

Poles.—The wires are supported at intervals of 125 feet or less on poles of wood, iron, or steel, illustrated in Fig. 57. Where the track runs close to the side of the street the pole shown at C may be employed, the wires being attached to insulating supports at the outer end of a side bracket which extends over the center of the track. Pole A may be employed with a double track in the center of the street; being provided with a double bracket for the wires. The method commonly preferred however is to erect two poles, like that shown at B, on opposite sides of the street, and connect them by a strong steel span wire or cable, from which the mains and feeders are suspended over the tracks on insulating supports.

Trolleys.—Connection between the motor and overhead conductor is usually made, on American roads, by means

ELECTRIC RAILWAYS AND RAILWAY MOTORS. 141

of the trolley, which is constructed with an arm, or pole, 12 to 14 feet in length, made of tough wood, or iron or steel

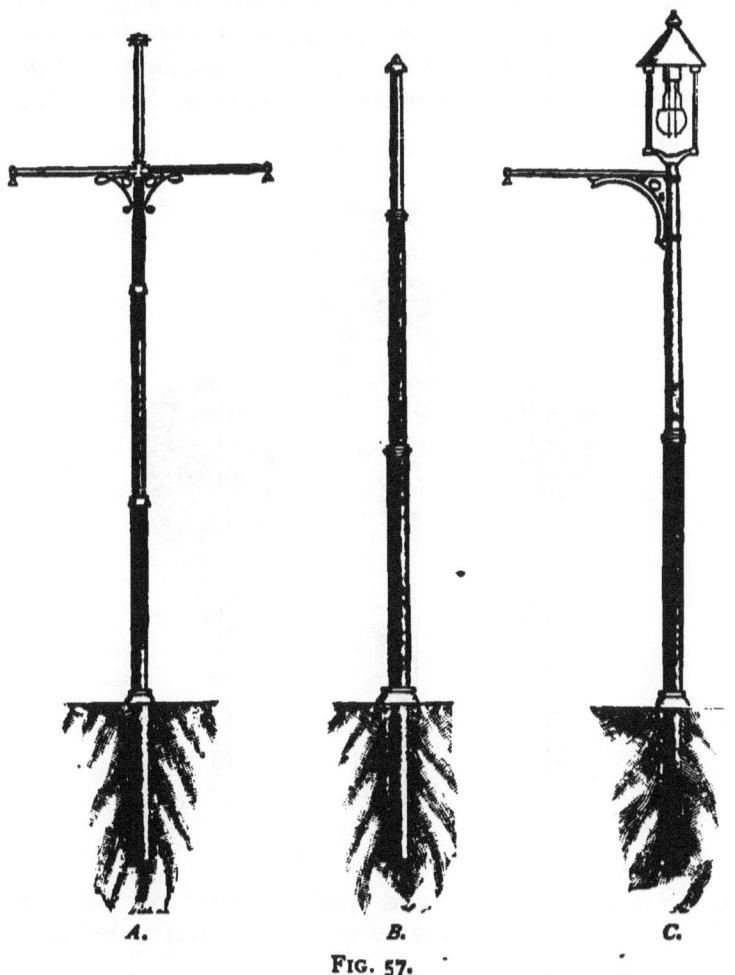

FIG. 57.

tube, carrying at its upper extremity a grooved bronze wheel, 4 to 6 inches in external diameter, with a groove $\frac{7}{8}$ of an inch to $1\frac{1}{2}$ inches in depth. This wheel, shown in Fig. 58, *A*, *B*, *C*, makes contact with the main conductor,

142 THE ELECTRIC TRANSFORMATION OF POWER.

or trolley wire, as it is usually called. It is mounted on a steel axle on which it rotates, and in its latest improved construction is furnished with graphite bushing, which requires no lubrication, is durable, and does not heat. Contact springs which press against it, as shown, insure continuous electric connection between the wheel and

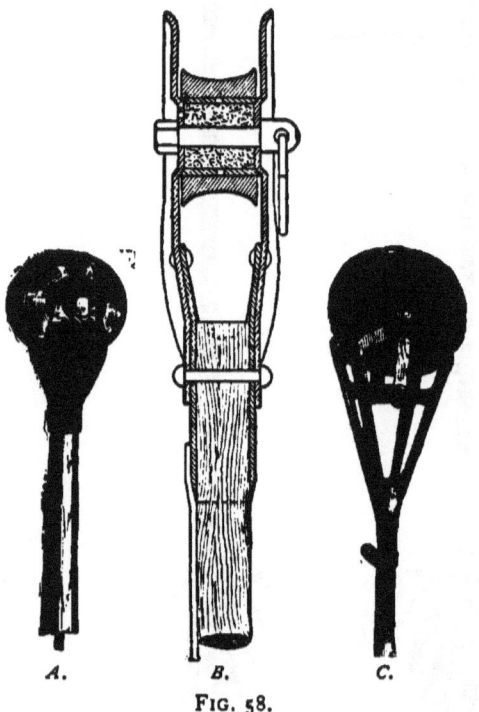

A.　　B.　　C.
Fig. 58.

motor through an insulated wire supported on the arm when the latter is made of wood, or attached to its base, when made of metal.

The wheel is also constructed without graphite bushing or contact springs, with electric connection through the axle, and oil lubrication. .

The arm is attached, at its base, to a socket mounted on a cast-iron stand which is bolted to the roof of the car, and furnished with springs which maintain it at an average elevation of about 45 degrees, and keep the trolley wheel in contact with the wire by upward pressure, while permitting a free lateral motion of the arm in rounding curves, or passing switches.

A cord attached to the upper end of the trolley arm extends down over the rear platform and gives the conductor control of the trolley for replacing the wheel on the wire, if it accidentally slips off, or making such other adjustment as may be required.

The direction of the arm, which is always toward the rear, can be reversed, at the end of the line, for the return trip.

The Boston Trolley.—This trolley, which is still in use on many roads, is illustrated by Fig. 59. It is constructed with eight spiral springs mounted horizontally on two transverse

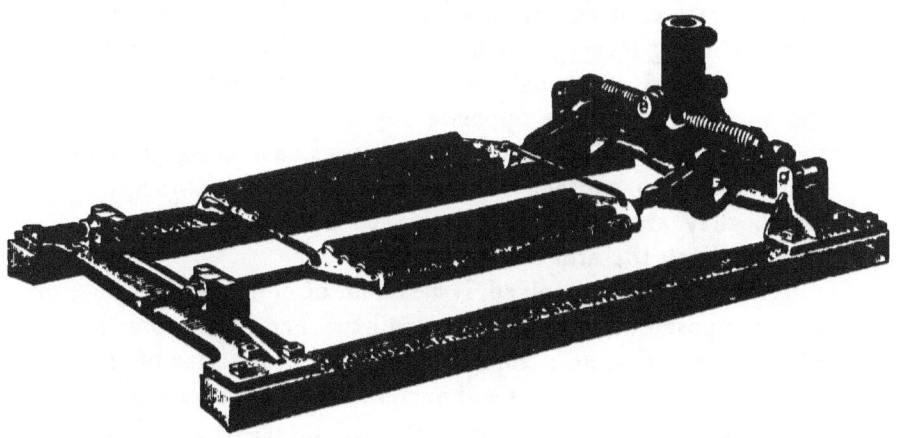

FIG. 59.

castings, as shown, and furnished, at the left, with screws by which the tension can be regulated, while, at the right, attachment is made to projecting supports on a transverse bar on which the trolley arm is mounted, which rotates in its bearings as the arm is depressed toward either end of the car; the leverage increasing the tension of the springs by their extension, and thus producing the upward pressure by which the wheel is kept in contact with the wire. A hinged support permits the required lateral motion of the arm, which is restrained and regulated by two transverse springs, as shown.

The Emmet Trolley.—This trolley, which is quite extensively used, is illustrated by Fig. 60. An iron base, bolted to the roof of the car, supports an iron frame which has a rotary motion on a central support, by which the requisite lateral motion is obtained. This frame carries four spiral springs, attached to hinged supports at each end, which are connected with a central support at the base of the trolley arm, on opposite sides, in such a manner that the leverage of the arm, as it is depressed in either direction, imparts corresponding leverage to one of the end supports, which rotates toward a vertical position, producing extension of the springs; while the opposite support is forced down at its outer end, bringing two projections, attached to its inner end, into contact with the side bars above them, by which they are firmly held; this support being the one toward which the arm is depressed.

This, it will be perceived, is a system of compound leverage, composed of a central lever and two end levers, one of which is in action at a time, in which the short arms of the central and end levers are connected. As the central lever moves toward a horizontal position, with increasing mechanical advantage, the end lever moves toward a vertical position with increasing resistance, produced by the greater tension of the springs attached to its longer arm. The mechanical advantage of the trolley arm varying in nearly the same ratio as the opposing resistance, produces an ap-

ELECTRIC RAILWAYS AND RAILWAY MOTORS. 145

Fig. 60.

proximately uniform pressure of the wheel against the wire at any angle of elevation of the arm.

The Compression Spring Trolley.—The construction of this trolley, which is very simple, will be readily understood from Fig. 61. Two strong spiral springs are supported, end

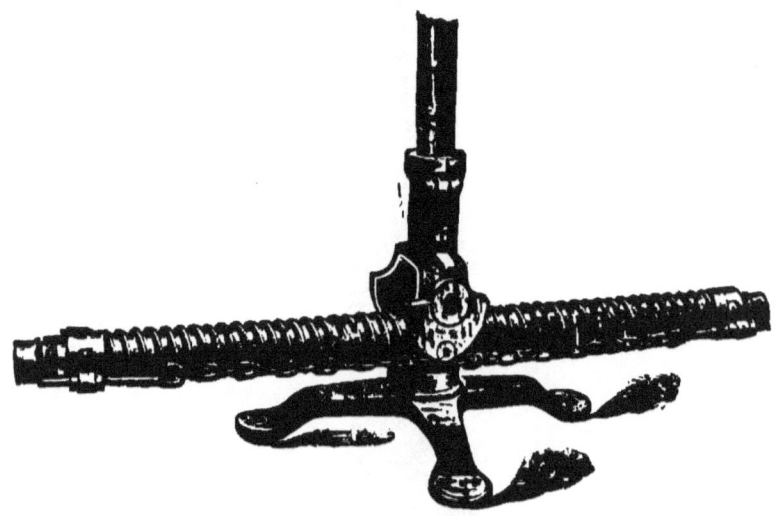

FIG. 61.

to end, on a 2½ inch iron pipe, 2½ feet long, capped at the ends, and pivoted at the centre on an iron base, so as to have a free rotary motion to provide for the lateral motion of the arm. The springs inclose the pipe loosely, permitting their free movement, and each is pressed against a central support by a collar which incloses its outer end and can move freely on the pipe. These two collars are connected together by two chains, one on each side of the pipe, each of which is attached, at its center to a semicircular projection which forms part of the base of the trolley arm. This base is mounted on a hinged support just above the pipe, and as the arm is depressed in either direction, the spring underneath it is compressed by the leverage, pro-

ELECTRIC RAILWAYS AND RAILWAY MOTORS. 147

ducing the upward pressure of the wheel against the arm ; the tension of the opposite spring being, by the same movement, relaxed, permitting its extension by which the chains are kept taut in front.

The Siemens-Halske Sliding Contact.—This apparatus, used extensively in Europe, takes the place of the trolley, and is constructed, as shown in Fig. 62, with a light iron

FIG. 62.

frame, supported transversely on the roof of the car ; its top contact-piece, made of gas-pipe, extending across the roof, and curving downward at each end.

This frame swings on a hinge toward either end of the car, as required, and has a counterpoise at bottom, by which the contact-piece is pressed upward against the line wire, the frame being inclined backward at any required angle. Hence it is impossible for the contact to be broken, even in rounding the shortest curve, as the contact-piece is long enough to permit the line wire to slide on it laterally to any required extent, in either direction, and the curved ends allow it to pass easily under a cross-line without risk of catching or breaking the wire.

The Tube and Piston Contact.—There are other methods of construction still in use on some of the older roads; among which may be mentioned the tube and piston method, formerly employed by the Siemens-Halske Co. in Europe, in which the line consists of an iron tube supported above the track, and slotted underneath; in which connection with the motor is maintained by a conducting arm, connected through the slot with a sliding piston which travels in the tube.

The Double Trolley.—The double trolley has also been employed to a limited extent; the circuit consisting of two line wires with a trolley in contact with each; the current flowing to the motor through one wire and returning through the other.

Insulators and Clamps.—The trolley wire is supported by clamps, or clips, attached to insulators, and the latter are made of different forms and have different modes of attachment adapted to span wires, brackets and curves. A common form of bracket insulator is shown in Fig. 63. It is made of hard rubber of special composition adapted to out door exposure, and is bell-shaped to shed the rain. Iron nuts are imbedded in it, into which the clamp which supports the wire is screwed below, and an eye-bolt above by which it is suspended from the bracket.

The clamp is usually made of bronze and carries the wire

in a groove on its under side; and is so formed as to offer the least possible obstruction to the passage of the trolley wheel, being curved underneath to conform to the groove of the wheel, and at the ends to prevent jar and insure proper contact as the wheel passes over. The wire may be soldered in the groove, or the lips of the clamp compressed round it, and may either be entirely inclosed, or a small section left exposed below for contact with the wheel; which is the preferable mode, though such contact is not absolutely essential, as the bronze clamp is a good conductor.

FIG. 63.

A span wire insulator, which hangs from a span wire between poles on each side of the track, is shown in Fig. 64, attached above to a steel support having hooks at the ends by which it is suspended on the span wire, which passes through a grooved projection at the center by which it is held in position. In connection with this is shown an improved form of clamp, made in two pieces, one of which has a tongue adapted to a groove in the other, as shown at the end. The grooved piece has a threaded shank at its center which screws into a collar fitted to a projection on the opposite piece, by which the tongue is forced down into its groove, pressing the two pieces together and clamping the wire firmly in a groove adapted to it underneath, leaving a small section exposed for contact with the wheel. The hinge joint above the nut permits movement of the wire due to expansion and contraction. The clamp can easily be attached or removed, or loosened for adjustment of the wire.

150 *THE ELECTRIC TRANSFORMATION OF POWER.*

On curves the trolley wire requires support at the center of the curve and on each branch; the branch supports being the same as on other parts of the line, while the cen-

FIG. 64.

tral, or corner support, must be adapted chiefly to a horizontal strain. This is provided for by the curve, or "pullover" bracket shown in Fig. 65, which is attached to the corner pole by a swivel joint which permits movement of the bracket in any direction required by the strain. The steel shank of the bracket is incased, at its inner end, in a hard rubber insulator, fitted to a collar at the joint, against which an enlarged end section of the insulator presses, as shown, sustaining the horizontal strain.

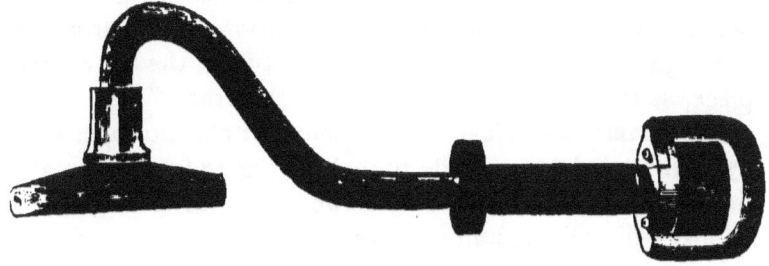

FIG. 65.

Switches.—A frog or switch is required in the trolley wire wherever one occurs in the track, and is similar in form to the latter but not usually movable ; and as it requires both strength and conductivity it is usually made of bronze, and is of different forms.

Three-way Switch.—Fig. 66 shows a three-way switch required where two branch lines are connected with the

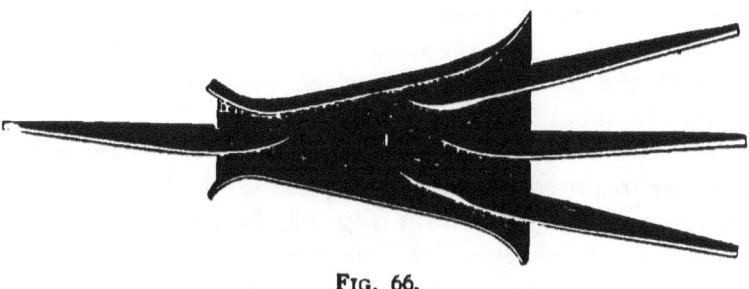

FIG. 66.

main line, at the same point. It is constructed with a flat central piece from which four flanges project, three to the right and one to the left, to which the wires are attached ; the main line to the left flange and central flange on the right, and the branch lines to the other two on the right.

As the trolley wheel leaves the wire it is received by the flange, and guided to the opposite flange by the movement of the car through the corresponding switch in the track, assisted by the conductor's manipulation of the trolley; the edges of the wheel being in contact with the flat centre piece, as it passes from one flange to the other.

The Emmet Switch.—The Emmet switch, shown in Fig. 67, is intended for a single branch diverging from the main line to the right or left, or for two branches diverging in opposite directions, and is accordingly made in three forms

FIG. 67.

with flanges in each, diverging in accordance with this arrangement. The central part is short, curved at the sides, and deeper toward the branches than the bottom line of the flanges, so that the wheel travels continuously on the edge of the flange, and is received by each flange as it leaves the opposite one.

The Atkinson Switch.—On single track roads where turnouts are required to allow the passage of the cars, the Atkinson switch, shown in Fig. 68, furnishes a reliable guide for the wheel, which does not require manipulation of the trolley by the conductor, as in other switches.

It is made of thin spring steel coated with zinc to prevent oxidation, and has two flanges, a straight one, and one curved vertically, as shown. The latter comes close to the former at the left, so that the trolley wheel of a car going from left to right enters both flanges on the left, and as it moves to the right is guided by the curved flange under the straight one. The trolley wheel of a car going in the opposite direction travels on the straight flange, and passes through, pressing the curved flange aside, which springs back to its place after the wheel has passed. One of these switches is required at each end of the turnout, their relative positions being reversed, so that, at one end, the curved flange guides the trolley wheel to the turnout, and at the other end, to the main line.

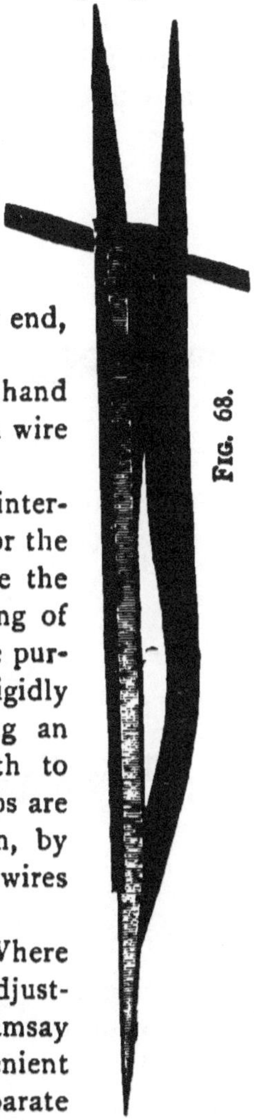

They are made for either right or left hand turnouts, and are supported on the span wire by the cross strap shown at the right.

Right-angled Crossing.—Where lines intersect each other, crossings are required for the guidance of the trolley wheel, and where the intersection is at right angles a crossing of the kind shown in Fig. 69 answers the purpose. It is made with four flanges rigidly attached to a center piece, and having an opening between their ends underneath to allow the passage of the wheel. Clamps are provided on the upper side, as shown, by which the crossing is supported on the wires by soldering or otherwise.

The Ramsay Adjustable Crossing.—Where lines intersect at oblique angles, an adjustable crossing is required, of which the Ramsay crossing, shown in Fig. 70, is a convenient form. It is constructed with two separate pairs of flanges which intersect each other, and are clamped together at the center by a bolt and washers at any required angle. The lower end of the bolt being on a line with the bottom of the flanges, at the central point, supports the

FIG. 68.

wheel in passing through the open space, and is shaped to conform to its curve. The trolley wires lie in clamps on the upper side under which are holes for supporting wires.

Where power is derived from independent sources, as on

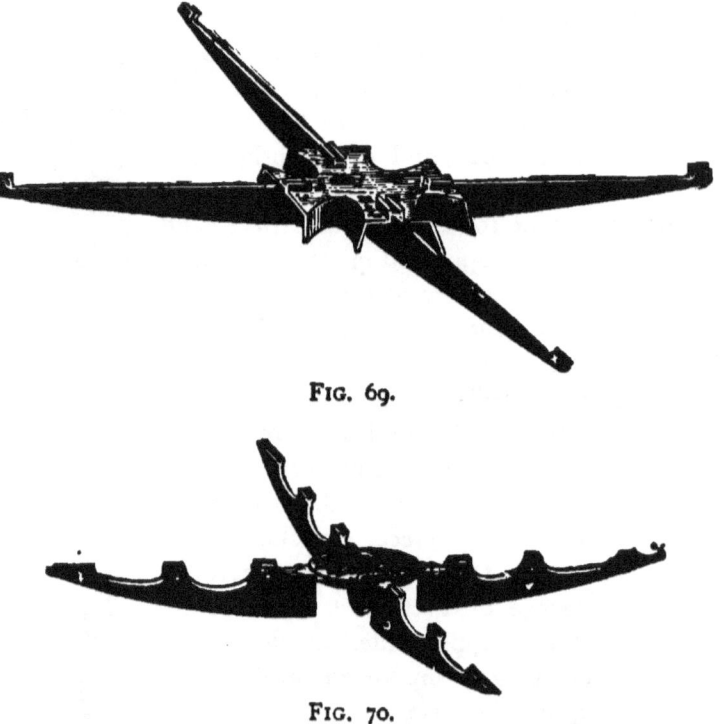

FIG. 69.

FIG. 70.

intersecting lines belonging to separate companies, adjustable crossings are required in which the two parts are insulated from each other, and the wires kept from contact at the point of intersection.

Trolley Line Breaker.—Lines constructed in insulated sections require insulating connectors, known as "trolley line breakers," for joining the sections together. Where these are supported on span wires or brackets they may be constructed with clamps insulated from each other to which the wires are attached and on which the trolley wheel travels ; but they may also be so made as not to require support. Fig. 71 shows a line breaker which combines lightness and strength, and may be placed at any convenient point on the line.

It is made with a double-headed steel bolt incased in hard rubber for insulation, and furnished at each end with links, fitted to it loosely, to which the wires are attached. Underneath this is a light wooden skate for the guidance of the trolley wheel, grooved above for the wires, and suspended from the bolt by a metal support. The momentum of the car, at ordinary speed, is sufficient to carry the wheel, without current, on this skate.

The Johnston Disconnector.—This simple device is designed is cut off the supply of current from any section of the line wire in which a break may occur, without stopping the supply through the feeder wires to the other sections ; so that the accidents liable to occur from contact with the ends of a broken electric wire may be averted, and repairs made without incurring similar risk.

It is constructed as shown in Fig. 72, with two horizontal arms attached to opposite sides of a wooden center-piece, in each of which is a spiral spring enclosing an iron rod connected with a hinged line connector at its outer end, and with a sliding catch, at its inner end, above which is a clamping switch, connected with it when the spring is under tension, as shown on the left, but disconnected and pulled upward by a spiral spring when the main spring is not under tension, as shown on the right. The two rods are connected through the switches by a wire, and the apparatus supported on an insulator, as shown.

When the line wire is unbroken its tension keeps the main springs extended, under tension, and the switches clamped, so that the line current traverses the apparatus ;

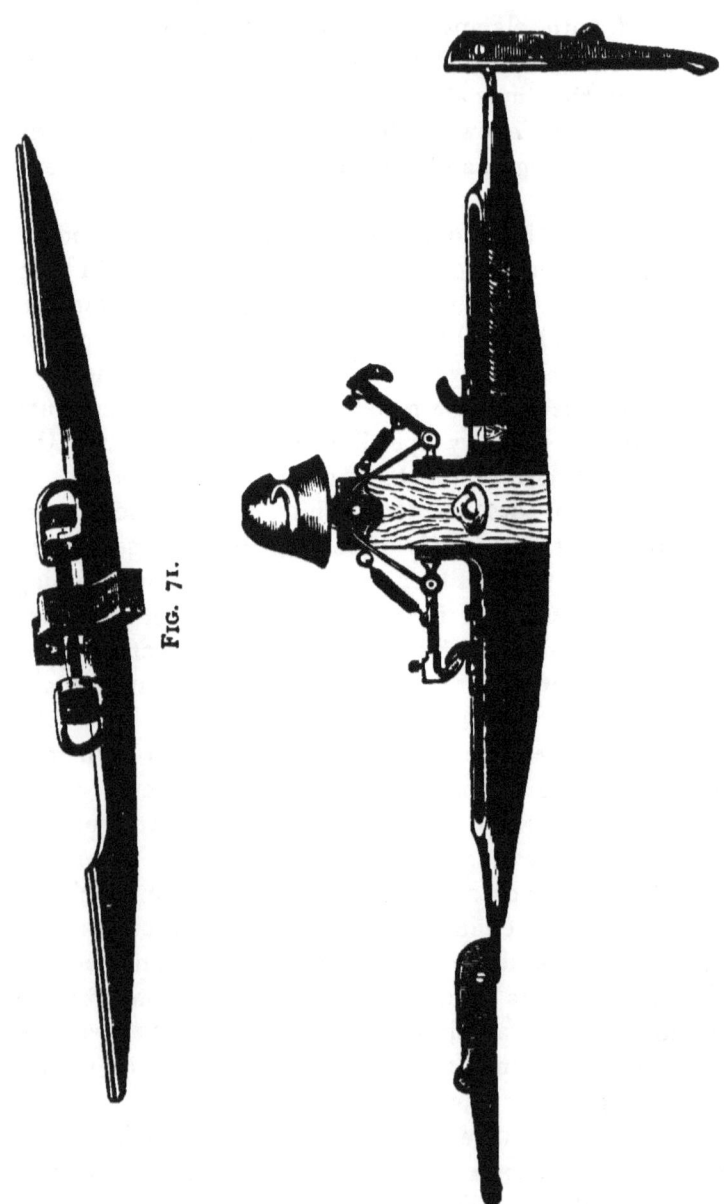

FIG. 71.   FIG. 72.

but when a break occurs, the tension being relaxed on each side of the break, the switch is released, the catch displaced, and the circuit opened as shown on the right; a similar result occurring at each end of the broken section, which is thus entirely cut out of the circuit, while the circuit in the adjacent sections, on either side, remains uninterrupted, as shown on the left.

Hence the ends of a broken wire normally carrying a large current, may, when the broken section is thus cut out, drop on a telegraph or telephone wire, or on a passing horse, vehicle, or person without producing electric injury; and, in case of fire, a section may be instantly cut out of the circuit by merely cutting the wire. The circuit remains open for the protection of the line men while repairing the break, and is closed, when the repairs are completed, by closing the switches.

These disconnectors require to be placed about one fifth of a mile apart, and can take the place of the ordinary insulated clamps to prevent incumbering the line. They can also be employed on the feeder wires and on electric light wires.

Railway Motors.—The work required of a street railway motor, and the conditions under which it is performed are essentially different from those of a stationary motor and require motors of special construction which have been developed by the progress of the electric railway system. The difference between the stationary motor and the railway motor is similar to that between the stationary steam engine and the locomotive engine. The railway motors are mounted on the truck, under the car, between the axles, both for economy of space and for direct connection; two being generally employed, geared to the two central axles, so that the truck and motors become an electric locomotive, in which the driving wheels are closely connected with the source of power, as in the steam locomotive. The motors are therefore exposed to snow, rain, mud, dust, sand, flying pebbles, dripping water, and contact with obstructions between the rails, which would soon render them useless unless amply protected. In addition to this the constant jarring and vibration of the rapidly moving truck, though reduced by spring mounting, requires greater strength of construction, in such a motor, than in one that is stationary.

The variation in electric strain is much greater than in stationary motors and requires to be provided for by employing wire of larger current carrying capacity for the coils, and constant manual regulation through a rheostat, to meet the ordinary conditions of street railway work, especially in our American cities, where excessive crowding of the cars is a daily occurrence at certain hours; since a heavily loaded car may, at one instant, while running on the level, be impelled with a minimum current, aided by its own momentum, and, at the next require maximum current to move it from a state of rest up a steep grade.

The series wound motor has been found better adapted to this work than the shunt wound, from the fact that both its field and armature currents are easily controlled by manual regulation through the same rheostat; and that the current, when temporarily interrupted by insulating material on the track, is at once reestablished, both in field and armature, when the obstruction is passed, and the proper counter E. M. F. maintained. While, in the shunt motor, according to Kapp, a momentary delay occurs in the reestablishment of the field current, during which there is an abnormal flow of current through the armature circuit before the generation of counter E. M. F. by the reestablishment of the field current can take place.

The motors employed in the early equipment of electric roads, and still in use on many of the older roads, were modified forms of stationary motors, having the different methods of winding usually found in such motors. Their principal characteristics were a long drum armature of comparatively small diameter, revolving at high speed in a comparatively weak magnetic field, and connected with the car axle by two sets of gears to produce the double reduction of speed necessary to adapt the rotation of the armature to that required by the car wheels; the armature shaft carrying a pinion which engaged a large gear on an intermediate shaft, which likewise carried a pinion engaging a large gear on the car axle.

The waste of power involved in the production of unnecessary speed and its subsequent reduction in this manner, requiring, for the two motors, eight gears and twelve shaft bearings, with their attendant friction, soon became apparent; the extra wear of gearing, bearings, brushes, and commutator, and extra consumption of oil, required by this system, forming a considerable item in the expense of maintenance; while the occupancy of space under the car made the proper mounting and equipment of brakes difficult. The noise of this complicated high speed gearing, and the continuous squeaking of the brushes on a high speed commutator, were also found to be highly objectionable.

The remedy was evidently to be found in reduction of the armature's speed without reduction of the motor's power. As the energy of the motor depends on the number of lines of force cut by the armature per unit of time, it became evident that by reversing the method of construction, replacing the long armature of small diameter, revolving in a weak field, by a short armature of large diameter revolving in a strong field, the same number of lines of force would be cut per unit of time, at slower speed, thus accomplishing reduction of speed without reduction of power; and on this principle, variously modified, railway motors are now constructed. They are also iron clad, being incased in a strong iron or steel armor which

160 *THE ELECTRIC TRANSFORMATION OF POWER.*

excludes all injurious matter and affords ample protection to every part, electrical and mechanical; and are con-

Fig. 73.

structed with special regard to economy of space and utility of material.

The reduction of armature speed, as above, has resulted in the elimination of the intermediate shaft and high speed gearing; single reduction of speed by a pinion on the armature shaft engaging a gear wheel on the car axle, taking the place of double reduction by two sets of gears; the speed of armature and gears being thus reduced one half, with corresponding reduction of wear on gears, bearings, commutator and brushes, and hence of expense of maintenance; the noise, so annoying to passengers, being also almost entirely eliminated.

The Westinghouse Single Reduction Railway Motor.—This motor is made with a circular, multipolar field, the construction of which will be readily understood from Fig. 73. To a strong, rectangular, iron frame, are hinged, at opposite ends, two semicircular iron yokes, one above and the other below, which are fitted to each other, and bolted together, forming a cylinder which incloses the armature and field coils and protects them from external injury. On the inside of this cylinder are cast four rectangular pole-pieces, at equal distances apart, which also constitute the cores of the field-magnets; and on these are placed the field coils, previously wound on a frame, and wrapped with an insulating covering; one of which is shown in Fig. 74. Each coil is secured in its place and protected by a heavy brass cap, through which the pole-piece protrudes, and which is bolted to the yoke; and by removing this cap the coil can be taken off for repairs or replacement by a new one. The four coils are connected together in series in such a manner as to produce alternate north and south poles.

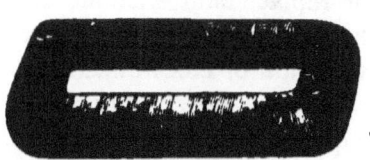

FIG. 74.

By distributing the coils on four cores instead of two, the quantity of wire in each is proportionally reduced and the radiation of heat increased; larger wire can be employed, having fewer convolutions in each coil, thus materially reducing the resistance and consequent heating and consumption of energy, and increasing the current carrying capacity; shorter cores are required, giving more ample space for an armature of large diameter, as required for reduction of speed; and by bringing the coils into closer proximity with the armature and pole-pieces, the magnetic resistance is reduced and magnetic efficiency proportionally increased.

The circular form of the motor gives not only maximum mechanical strength and economy of space, but prevents the dissipation of magnetic energy from projecting angles, confining it to the interior where it is required for the performance of work, as already referred to in connection with stationary motors.

The armature is of the Siemens, or drum, type, constructed with a laminated core mounted on a steel shaft, and grooved on the surface for the reception of the coils, which are wound to fit the grooves, driven into place, and subsequently connected in series with each other and with the commutator. Hence any coil can be removed for repairs or replacement without disturbing the others; and, being firmly imbedded in the iron, it cannot be displaced by vibration or electromagnetic action. They are constructed with wire of comparatively large gauge, whose carrying capacity is equal to the maximum electric strain which may be required of it, and hence they are not liable to heating or burning out. The close proximity of the core to the pole-pieces and of the armature coils to the field coils reduces the resistance to the minimum and gives maximum electromagnetic efficiency.

The commutator is substantially constructed with special reference to street car work; its segments rest on a bearing surface throughout their entire length, so that they cannot be bent by any accidental external force, and provision is made against their unequal expansion by heat and consequent loosening.

Carbon brushes are employed, set radially opposite the centers of the two upper pole-pieces, and hence making contact at the extremities of a quarter arc on the upper surface of the commutator, the chord of which coincides with the neutral line in this construction of the field. They are supported in this position by the brush-holders shown in Fig. 75, in which they are held by two spring clamps attached to insulating oak arms,

FIG. 75.

supported on a casting of composition metal, bolted to the motor frame under the commutator.

The frame maintains parallelism between the armature shaft and car axle, so that the gears mesh together properly. The armature may be removed when required, by loosening the lower yoke and swinging it down on the left hinge underneath, into a pit. In the frame are shown the oil boxes, above the bearings, which are provided with metal caps, as shown in the next cut.

Each car has two motors, mounted on the truck as shown in Fig. 76; each supported at one end on the car axle, and at the other on spiral springs resting on cross-beams; their relative positions being reversed. In the left hand motor is shown the commutator and brushes above, and the gearing of the armature shaft to the car axle, below. These parts are protected, in both motors, by iron coverings, as shown in the right hand motor; the gears being inclosed in a separate case, which also suppresses the noise, rendering them practically noiseless. The coverings of the other parts extend across the frame, above and below, and vertical end coverings are also provided, which may be attached to the ends

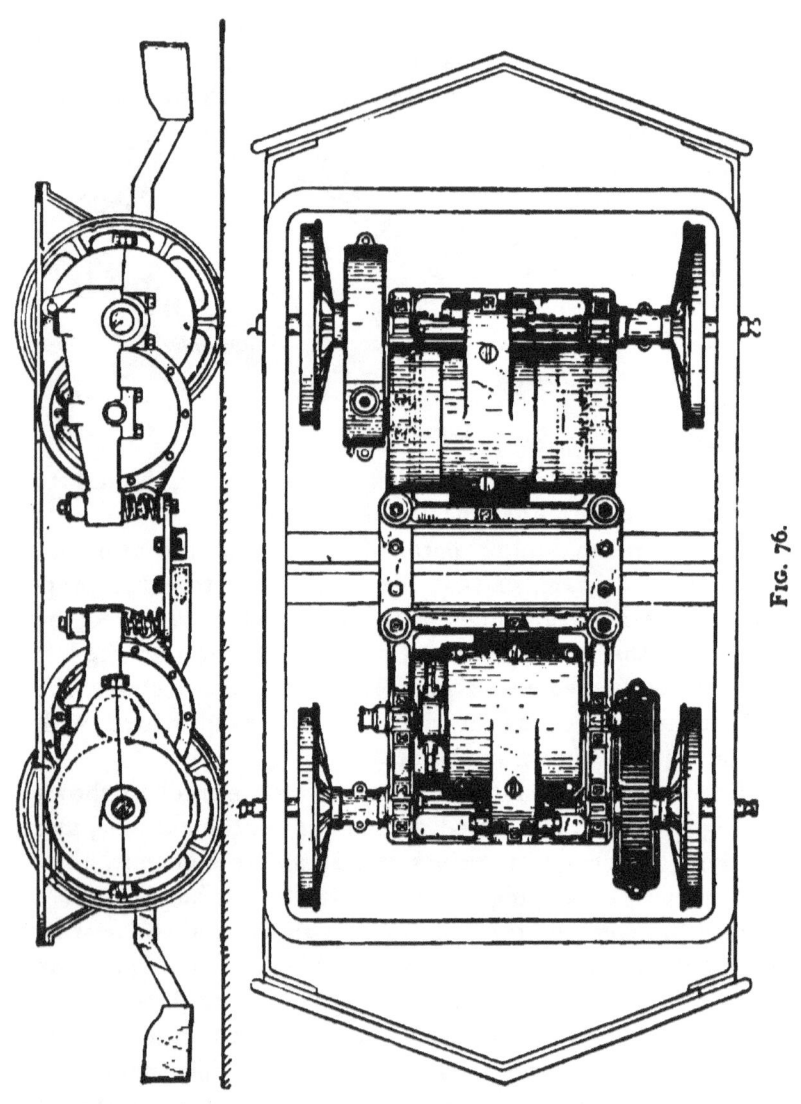

FIG. 76.

of the yokes if required, sufficient opening being left for ventilation.

Controller.—On each platform of the car is a controlling stand, made with an iron case having the form of a half cylinder, as shown in Fig. 77, covered with a water-tight cap, and mounted vertically with its flat side next the dash-board. Its internal construction, shown in Fig. 78, is as follows:—At the center of the convex covering in front is a vertical support, on the inner surface of which are mounted, with springs, ten brass stops, shown by the circles, to which the terminals of the electric circuit are attached. Just behind this support is a vertical switch, which carries eight convex, brass contact blocks, three on the right and five on the left of a central vertical space, as shown. The two bottom blocks, A and D, are in the form of steps, receding each way from the bottom upward, the bottom steps twice the width of the others. Above these on the left are four rectangular blocks, E, F, G, and H, opposite which, on the right, are two, B and C, each as wide as two of the left hand blocks.

FIG. 77.

This switch can be rotated by the handle shown in Fig. 77, in either direction from the projecting stop on the rear of the cover, with which a flange on the handle makes contact; bringing the blocks into contact with the stops, and thus closing the circuit, which is open, as shown in Fig. 78, when the handle is at the front central position.

Controller Connections.—The connections and course of the electric circuit are as follows: A rheostat, or "current diverter," as it is called in this case, is placed under the car and connected with the controller. It is composed of four iron wire resistance coils, contained in a box and con-

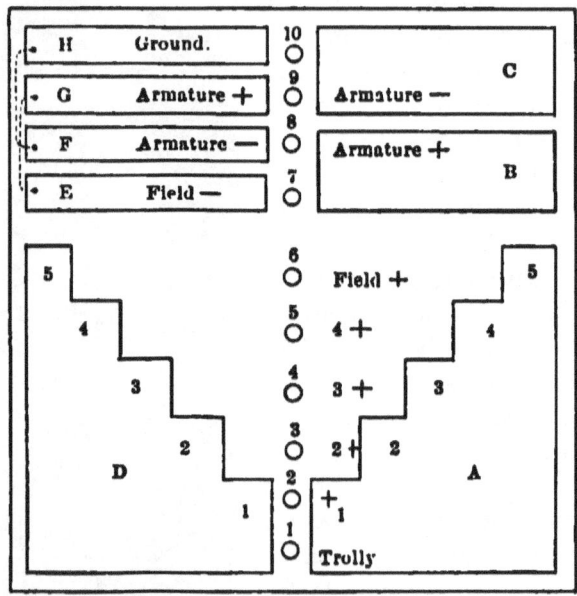

FIG. 78.

nected in series with each other and with the field coils of the motor. The terminals of its coils are connected with the brass stops, 2, 3, 4, 5 of the controller; stop 1 is connected with the trolley, stops 6 and 7, with the field, 8 and 9, with the armature, and 10 with the rails through the car axle and wheels. When the switch is turned from right to left, in the direction in which watch hands move, blocks *A*, *B* and *C* make contact with stops 1, 2, 7, 8, 9, 10; current from the trolley enters by stop 1, passes through the lower step of block *A* to stop 2, thence through all the diverter coils in series and back to stop 6, thence through the field coils and back to stop 7, through block *B* to stop 8, thence through the armature coils and back to stop 9, through block *C* to stop 10, and thence to the rails.

As the switch is turned farther, step 2 is brought into contact with stop 3, cutting out the resistance coil connected with stop 2, the current taking the shorter course through stop 3; in like manner as steps 3 and 4 are successively brought into contact with stops 4 and 5, the coils connected with these stops are cut out; and when step 5 is brought into contact with stop 6, all the resistance coils are cut out and the full current passes direct to the field.

When it becomes necessary to reverse the rotation of the motors for any purpose, such as backing, or stopping promptly in case of danger, it can be done by turning the switch from left to right, from the neutral position, bringing the five left hand blocks into contact with the stops. Block *E* being connected with block *G*, as shown, *F* with *H*, and *F* and *G* with the armature through the opposite brushes, the field current from stop 7 passes to the armature by way of blocks *E*, *G*, and stop 9, returns through stop 8 to block *F*, and passes through blocks *F*, *H*, and stop 10 to the rails. The armature current being therefore reversed, as may be seen by comparing the signs on blocks *F* and *G* with those on blocks *B* and *C*, while the direction of the field current remains unchanged, the rotation of the motors is reversed.

Both motors are connected together in parallel, on circuits connecting the wire leading from the trolley with the wire leading to the rails. The controller on each platform is connected with the main circuit leading to both motors, so that the current can be controlled from either. The motor man always stands on the front platform, and, at the end of each trip, leaves the controller he has been using in the neutral position, and uses the one on the other platform during the return trip. The handle of the controller is connected with a ratchet-wheel having a notch for each of the five positions required in admitting current through the diverter, as described ; so that the different positions can be distinguished by the difference of resistance and click of the ratchet-wheel while the motor man keeps a look-out ahead, as required.

Each motor is connected with a separate diverter, at opposite ends of the car, and a separate cut-out at the center where each wire passes through a connection with the main circuit, which can be opened by a switch connected with a handle under the car seat, so as to cut the motor out of the circuit when required for any purpose, as when testing for the location of faults. The main circuit has a branch connection with a lightning arrester, connected with the ground through the car axle, wheels, and rails ; through which, in case of the line being struck by lightning, the charge is diverted from the motors and other apparatus, taking the shorter course, either downward or upward, in preference to the longer one through the apparatus, on account of the enormous potential difference between the earth and cloud.

Lightning Arrester.—The lightning arrester is shown in Fig. 79. It is constructed with an air-tight case which incloses, at its base, two carbon blocks, one on each side, each connected above with a vertical row of carbon points which approach within $\frac{1}{78}$ of an inch of each other at the center. On each side of the case outside is a brass arm, mounted on an insulating marble slab, as shown on the right, one of which is connected with the trolley wire and the other with the ground wire by binding screws at their upper ends. Each is hinged above, as shown, and terminates below in a curved horizontal brass arm, with a carbon tip, which passes loosely through a hole in the slab and makes contact with the carbon block inside, thus completing the circuit through the points. On top of the case is mounted, on a vertical standard, a cross-bar having insulating blocks of hard-rubber attached to its opposite ends,

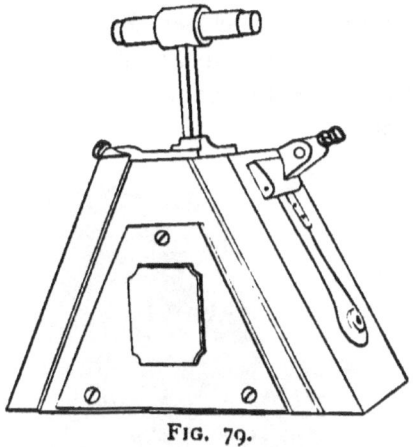

FIG. 79.

against which the lower ends of the arms strike when thrown upward.

When lightning strikes the line, the charge passes in by one arm and out by the other, through the carbon blocks and across the points. The carbon becoming heated, a momentary arc light is produced between the points, and the air, suddenly expanded by the heat, drives out the arms, thus breaking the circuit and extinguishing the arc. The arms are thrown up against the insulating stops, and instantly drop again into their places, closing the circuit in readiness for the next charge.

Carbon points and contacts are preferable to metal ones for this purpose, which are liable to be melted by the intense heat, while the carbon is insoluble by heat, and is not materially wasted by combustion during the momentary duration of the arc.

This arrester is inclosed in an iron box, for its protection,

170 *THE ELECTRIC TRANSFORMATION OF POWER.*

when employed on cars; when employed in stations, the box is not required. The front slab is removable for inspection of the interior, adjustment of the inclosed parts, and repairs. The joints at the edges of this slab, and at the arm-holes are made air-tight with soft-rubber packing.

The Thomson-Houston Water-Proof Single Reduction Motor.—This motor, shown in Fig. 80, is constructed with a ring armature of the Pacinnoti-Gramme type, $8\frac{1}{2}$ inches wide and 20 inches in diameter; the ring being $5\frac{1}{4}$ inches in

FIG. 80.

thickness. It has a laminated core, mounted in the usual manner on a steel shaft. The coils are wound by hand, with No. 12 wire, between T-shaped flanges which extend across the surface of the core; each of which is $\frac{3}{8}$ of an inch thick, $\frac{3}{4}$ of an inch high, and has a lip on each edge above, which increases the width of the upper surface to $\frac{11}{16}$ of an inch. The coil space between each pair of flanges is $\frac{9}{16}$ of an inch wide below, and $\frac{1}{4}$ of an inch between the lips above. The terminals of adjacent coils are electrically welded together, instead of being soldered in the usual manner. Each coil is about ½ an inch in thickness, leaving a quarter inch space between its upper surface and the lips of the flanges, into which a wooden strip is driven, by which the coil is firmly wedged down in its place, so that no binding wire is required. The armature is therefore practically iron clad, the coils being thus completely protected, and almost the entire surface brought into close proximity to the pole-pieces of the field; the space required for clearance being much narrower than would be safe with the coils wholly or partially exposed in the usual manner, and hence the magnetic resistance of this space is reduced to the minimum.

The motor is inclosed in a strong protecting case, made of soft cast-steel, which also serves as its supporting frame, and becomes part of the magnetic circuit. This case is divided into two sections, hinged above and below at the car axle, on which its outer end is supported; so that it can be opened either from above or below, as may be found most convenient, for the inspection or removal of the inclosed apparatus. Its inner end rests on a central spring support. The lower section is shaped externally, at each end, like the bow of a boat, with a prow to throw aside obstructions; its bottom being flat and its sides convex. It can pass through water up to the car axle without wetting the inclosed apparatus. The armature shaft rests on bearings at the sides of the case, provided with caps on which are mounted the grease boxes. It is fitted with steel shells which rotate in bushings of soft sheet metal, both of which can be cheaply replaced when worn.

The upper section of the case is made with openings at the sides for ventilation, and carries the field coil shown at A, the only one employed, which is wound on a gun-metal frame bolted to the case above; No. 4 wire being employed for motors of slow speed and No. 7 for those of higher speed. This coil with its supporting frame is 2½ inches thick and 3¾ inches in vertical height; and surrounds the upper part of the armature when the case is closed, ample space being left for clearance.

On each section of the case is cast a pole-piece, which projects from its interior surface; the upper one being shown at B, and the lower one occupying the corresponding position under the armature. They are of the same width as the armature, and approach within about 12 inches of each other at opposite ends; and between the upper one and the field coil is a space of about 1¾ inches.

The single field coil, surrounding the upper part of the armature, exerts on it an upward solenoid pull, so proportioned that, at normal load, it counteracts the force of gravity, relieving the lower bearings of the armature's weight, and thus materially reducing their wear.

The brush-holders are mounted on insulating blocks of wood, bolted in slots on the rim of the lower section of the case; and the brushes, which are of carbon, make contact with the commutator at angles of 60° with a radial line, thus preventing the disagreeable squeaking which brushes set radially are liable to produce, without interfering with reversal of rotation.

The gears are inclosed in an oil-tight steel case, which excludes dust and grit. It is made in two sections, and provided with a lid which can be opened for inspection and the introduction of grease, and a small cap which can be taken off for the removal of the pinion. It is kept filled with grease, which is reduced to a semi-fluid state by the heat generated by friction. The gears, thus lubricated and inclosed, are rendered practically noiseless, and their wear greatly reduced.

This motor is so compactly constructed, that it can be employed on roads of three feet gauge as well as on those of standard gauge, and two can be mounted on trucks having only a five-foot space between the axles.

The motor circuit and connections, including controllers, rheostats, cut-outs, and lightning arrester, are constructed on the same general principles as those already described, so that a detailed description is unnecessary. The motor is series wound, and the two employed on each car are connected together in parallel; the armature and field connections being made at the controlling stands on each platform.

The Curtis Single Reduction Box Motor.—The construction of this motor is illustrated by Fig. 81. Among its chief characteristics are compactness, ample protection, and ease of access for inspection or repairs. The case is a rectangular box 24 inches square and 21 inches deep, made of soft steel ¾ of an inch thick, closed at the sides and bottom, and water-tight up to the car axle; and connected with it, as parts of the same casting, are the lower half of the gear case with its axle bearing, the bearings of the armature shaft, and the cores of the field-magnets and lower halves of the pole-pieces.

The field-magnets are mounted vertically on each side of the armature, as shown, and the case becomes the connecting yoke between them, forming part of the magnetic circuit. Their coils are kept in place by clamping plates bolted to the case, and are easily removed for repairs by loosening the bolts. The pole-pieces, which curve inward in the usual manner, have their upper tips removable for the reception or removal of the armature, and are secured in place with bolts. There are also two rectangular side pieces fitted closely to openings above the armature bearings, so as to form part of the case, which are removable for the same purpose, and have the bearing caps cast in connection with them.

The armature is of the Gramme ring type, with laminated

174 *THE ELECTRIC TRANSFORMATION OF POWER.*

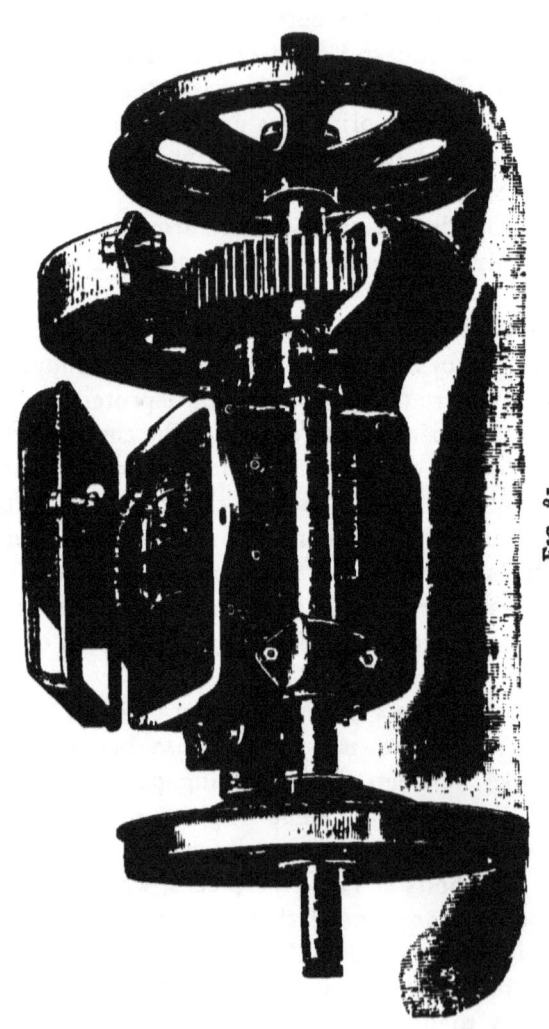

FIG. 81.

core, grooved for the reception of the coils; and having insulating troughs of molded mica fitted to the grooves, in which the coils are wound. The coils are composed of flat copper wire, or tape, which, being more flexible than round wire, is more easily wound and occupies less space; and they are of ample carrying capacity for the largest currents required. They are confined in the grooves by strips of hard insulating material above them, firmly secured by transverse metal bands, sunk in grooves below the surface of the core projections. Their connections with the commutator are made with flexible wire wound with a spiral turn of about 90 degrees, so that the brushes can be placed on a horizontal line, approximately at right angles to the neutral line, to be easy of access. The joints are all wound with fine copper tape, no solder being employed, and are therefore easily repaired.

The armature coils are connected in series with the field-magnet coils; and their combined resistance is about one ohm. Only 190 pounds of copper are employed for both.

The commutator is inclosed within the case, and is of special construction which improves its action and causes it to maintain a smooth surface; increasing the durability of the brushes, which are of carbon set radially, and rendering their action practically noiseless.

The case has a light cast-iron cover, provided with a large opening through which the commutator can be inspected and the brushes renewed, and smaller ones for ventilation; the large opening being covered with a canvas flap to exclude the dust. Access to the motor is obtained through a trap door in the car floor; and when repairs are required, the cover can be removed, also the side pieces and pole-piece tips, and the armature taken out; thus avoiding the usual troublesome method of removing it from below, with the car placed over a pit.

The interior of the motor is amply protected from mud,

water, and dust, not only by the inclosing case, but also by thick water-proof canvas coverings over the field coils and exposed ends of the armature coils.

The lower part of the gear case being cast integral with the motor case, cannot become loosened or detached. The gearing is designed to produce one revolution of the car wheel to four of the armature; giving an average car speed of about 15 miles an hour.

The motor weighs 2200 pounds, and is adapted to car wheels 30 inches in diameter; giving $4\frac{1}{2}$ inches clearance above the track.

Gearless Motors.—The superior advantages of single reduction over double reduction motors, for street railway service, having been amply demonstrated by practical experience, which has led to their general adoption, the next important advance in the same direction has been such further direct reduction of motor speed as to render the use of gearing for its reduction unnecessary. This has been accomplished by a further increase in the diameter of the armature, the quantity of wire and number of convolutions in its coils, and the strength of the field; all of which has been found to be within practical limits, with proper construction, resulting in the production of the gearless motor.

This motor has not yet come into general use, though it may be regarded as having passed the experimental stage of development. Its most earnest promoter in this country is Professor Short, whose motor, having undergone numerous modifications and improvements, may be selected as one of the best representatives of its class.

The Short Gearless Motor.—This motor, in its latest construction, shown in Fig. 82, is made with a triangular steel case having two sections, upper and lower; the upper section open at the sides for ventilation, and the lower, which is bolted and hinged to it, inclosed so as to be water-tight.

It has three field-magnets, distributed at equal distances

ELECTRIC RAILWAYS AND RAILWAY MOTORS. 177

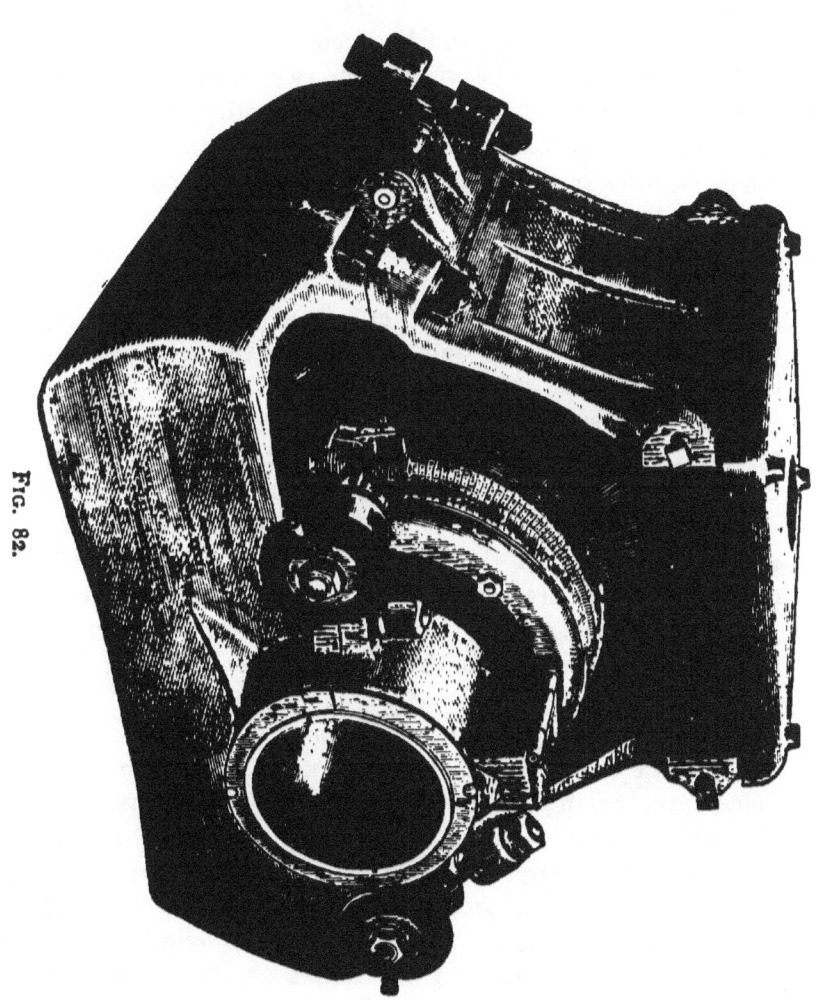

FIG. 82.

apart on the interior of the case, with pole-pieces projecting toward the face of the armature, as shown; their coils being connected together in series. These three poles are similar, being either positive or negative according to the direction of the rotation, and between each two magnets there is therefore a consequent pole of opposite polarity, whose pole-piece forms a part of the case, as shown. Hence there are six poles, three positive and three negative, with pole-pieces inclosing the armature.

The armature is 21 inches in diameter and 14 inches wide, and its core is built up in the usual manner with rings of soft sheet-steel, clamped together, whose planes are parallel to the plane of rotation, and grooved on the face, parallel to the shaft for the reception of the coils.

It is wound with 92 coils of copper wire, each having 16 convolutions, connected with 184 bars of the commutator, two to each coil; and is so cross-connected, that as the armature rotates in the six-pole field, half the coils are constantly in series and the two halves in parallel. Hence only two brushes are required, which are placed on a horizontal line to be easy of access; the flexible connections with the coils being arranged at the proper angle for this purpose.

It is mounted on a hollow shaft which is concentric with the car axle and is 6 inches in internal diameter, the axle being 4 inches in diameter, leaving 1 inch space between the two. This shaft rotates in bearings in the motor case, and the latter is supported on spiral springs on the truck by four supports, at the proper height to keep the axle and motor shaft concentric. Hence the entire weight of the motors being symmetrically distributed on the truck, with spring support, the rails are relieved from the excessive wear caused by the pounding of the wheels on their ends, in passing over joints, as in cars with geared motors which have direct support, in part, without springs, on the axle.

The apparatus by which motion is communicated to the car axle is shown in Fig. 83, and consists of a disk keyed to the axle next the wheel, from the rim of which two

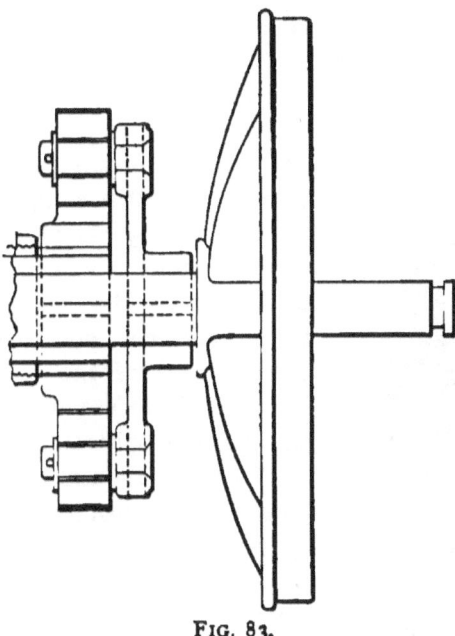

FIG. 83.

bolts project inward and are inclosed in rubber cylinders, each 4 inches long and 5 inches in diameter, which have bearings in a cross-bar slotted at the ends to receive them, and keyed to the armature shaft at the end opposite to that on which the commutator is mounted.

The two motors on the car take current in parallel, but may also be connected together in series when specially required. With the parallel connection they are designed for a speed of about 12 miles an hour, with full load, or 20 miles with ordinary load, on a level track; and with the series connection, for a speed of about 10 miles, with ordinary load.

The electric energy of the motor here described is 20

horse-power, its electromotive force 500 volts, its armature speed 120 revolutions per minute, its average efficiency about 83 per cent of the power derived from the steam engine, and its weight 2300 pounds.

Electric Lighting of Cars.—The lighting of electric cars is accomplished in a very simple and satisfactory manner by means of incandescent lamps, supplied with current on a circuit parallel with that which supplies the motors, having its own switches, and direct connection with the trolley and ground without passing through the rheostats or controllers, so that the stoppage of the motor current does not interfere with that which supplies the light.

Fig. 84.

The lamps require to be of special construction, with filaments adapted to resist the jarring of the cars, which would produce constant vibration of the long slender filaments in common use, giving a flickering light, and soon causing their rupture. One method of construction is to support the filament, at its bend, by a platinum hook imbedded in the glass at the point where the bulb is hermetically sealed, as shown in Fig. 84. Another method is to connect two short filaments, inclosed in the same bulb, in series, thus obtaining the same resistance and quantity of light as with one filament of twice the length, without the vibration and liability to rupture. Filaments of the ordinary length, without the hook support, but of special toughness and rigidity are also employed.

The advantages of such a system of lighting, in convenience, economy, and cleanliness cannot be too highly appreciated. The lamp is always ready, requiring no filling or trimming, and free from the well known annoyances of the oil lamp; and in case of extinction by the rupture of a filament, can be refitted with a new bulb, at a trifling expense, and in less time than is required to put a chimney on an oil lamp. Wiring the cars for this purpose, and mounting the lamps is cheaply and easily done, and the supply of current requires only a little more power at the generating station, in addition to that required to furnish current to the motors, the extra cost of which is comparatively small.

This system of lighting is also employed on steam cars; an engine in the baggage car, supplied with steam from the boiler of the locomotive, operating a dynamo which furnishes current to the lamps, either directly or through the intervention of a storage battery, which is constantly receiving current from the dynamo, and furnishing it, as required to the lamps; accumulating a sufficient excess of energy during the day to sustain the lamp current during the night.

Electric Heating of Cars.—The heating of street cars is an exceedingly difficult matter especially under the system of excessive crowding prevalent in our American cities. The coal stove, under the floor and seat, is perhaps the least objectionable of any of the ordinary methods; but with this the heat is not properly diffused and cannot be properly regulated; the fire is liable to be neglected, and the pipe is always in the way.

The heating of electric cars can be accomplished in a very satisfactory manner with electric heaters which are free from all these objections. Like the electric lamp, they are always ready for use, are placed under the seats, where they are entirely out of the way, and distributed in different parts of the car, so as to produce an even temperature, which can be so regulated and controlled as to keep the car comfortable, not only in severe winter weather, but in the chilly weather of spring and fall, when excess of heat makes the troublesome stove practically unserviceable. They are free from dirt, smoke, gas, and disagreeable odor; do not burn or scorch the woodwork or upholstery of the cars, or the clothing of the passengers, require no unsightly pipes, or pipe holes through the seats or roof, and are more economical than any other heaters. The disagreeable products of combustion pass off through the tall chimney at the generating station, and the heat is furnished on the cars by the simple movement of a switch. The electric equipment of a car is thus completed in a most satisfactory manner when it is heated as well as lighted by the same power which propels it.

The practicability of this system of heating has been tested by electricians for the last three years, during which numerous methods have been introduced and improvements made, and it may now be considered as having passed the experimental stage, having come into practical use on a large number of electric roads, where its efficiency and economy have been fully demonstrated.

The entire process except in the rare instances where electricity is generated by water power, involves the transformation of heat into power and of power into electricity at the generating station, and of electricity into heat on the cars. Notwithstanding these various tranformations, it is found that electric cars can be heated more economically in this way than by the direct consumption of coal; since, as in the electric lighting of such cars, the apparatus for supplying the electric energy is already in use, and only the employment of a little additional power is required. Careful estimates, made from electric heaters in practical use, show that the cost of heating a car in this way is less than half that of heating it with coal stoves; the average cost of electric heating, in two instances given, being 11 mills an hour, while that of stove heating, on the same cars, was 24 mills an hour.

The generation of electric heat, like that of electric light, depends on the passage of electricity through a conductor of high resistance, and the various heaters, while differing in details of construction, are all made on this general principle.

The Burton Electric Heater.—Prominent among those which have come into practical use, is the Burton heater, which may be selected as one of the best representatives of this class of apparatus. It is made, as shown in Fig. 85,

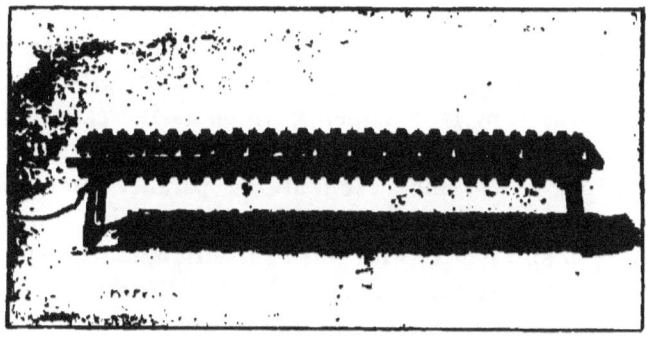

FIG. 85.

with a cast-iron case, 27 inches long, 8 inches wide, and 2 inches thick, including the projections with which it is studded above and below to increase its radiating surface. This case is mounted on iron feet, 4 inches above the floor, and is constructed with two plates bolted together, inclosing a space $\frac{1}{8}$ of an inch thick, closed at the edges with asbestos packing, in which are 75 feet of No. 26 German silver wire, having a resistance of $41\frac{3}{4}$ ohms, wound transversely on a light insulating frame, and imbedded in finely powdered dry fire clay, which absorbs and stores the heat, and prevents the oxidation of the wire.

Four heaters are required for a street car of ordinary medium size, distributed and connected as shown in Fig. 86, from which it can be seen, that by moving a switch to the left, the current must traverse the four in series, while by moving it to the right, the current passes from the line to the upper end of the car, where it divides and returns in two parallel circuits, each traversing two heaters in series. The latter connection, having only one-fourth the resistance of the former, is employed in heating the car rapidly, preparatory to starting, after which the full series connection is employed to sustain the acquired temperature; 30 minutes being the limit of safety for employment of the multiple series connection without oxidation of the wire. The current required on a 500 volt circuit, with the full series connection is 3 amperes, which is increased to 12 amperes during the few minutes the heaters are connected in multiple series, since each of the parallel circuits has only half the resistance of the full circuit, being only half the length, and hence, with this connection, carries current of twice the strength, 6 amperes, and therefore 12 for the two. Connection between heaters is made with No. 10 or 12 insulated copper wire.

By setting the heaters in tin cases, open in front, the heat is reflected into the car, and its diffusion under the seat, where it is not required, prevented. Seats paneled in front must be provided with openings opposite the heaters, to which wire screens may be fitted.

The success of electric heating on electric railways has directed attention to its advantages for steam railroads also, on which the cars can be electrically heated with the same facility as they are electrically lighted, as already described; current being generated on the train. If its efficiency and economy, on roads where it has already been introduced,

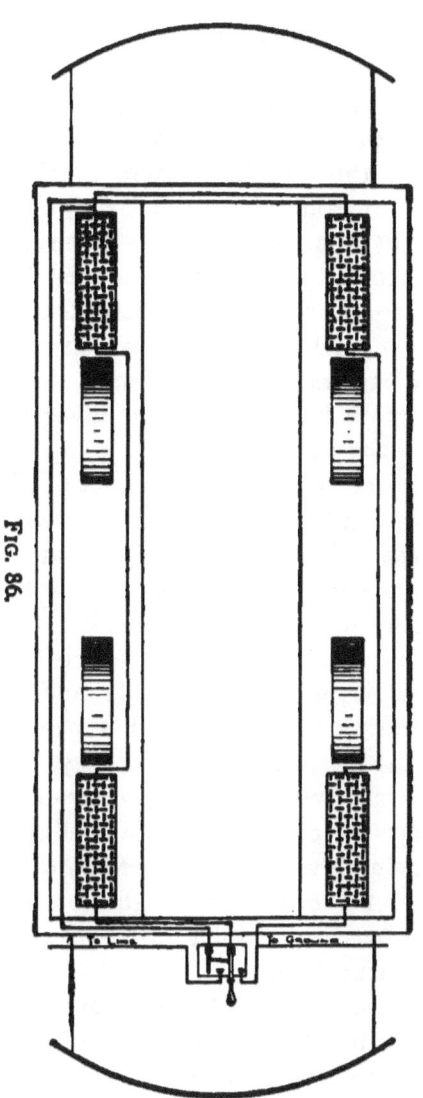

Fig. 86.

shall be found equal or superior to that of steam heating, its general adoption must soon follow, leading to the entire displacement of the coal stove, which has been found so dangerous in cases of accident.

The electric heating of buildings, and the employment of water-power for this purpose, where available, is also in contemplation but cannot be appropriately considered here.

The Conduit System.—The success of electric street railways constructed on the overhead system has stimulated electric inventors to produce a practical conduit system, and the progress made within the last three years, as the nature of the obstacles and the means of overcoming them have become more fully understood, gives hope of complete success in the near future.

The difficulty of maintaining contact with the electric conductors in a conduit, to which allusion has already been made, is perhaps full as great as that of maintaining proper insulation. When the trolley slips off an overhead wire, as it sometimes does in rounding a curve, it is but a moment's work to replace it again, and no damage is done, but when it slips off a conduit wire, it is not only much more difficult to replace, requiring the opening of a man-hole to which the car must be pushed, but the trolley is liable to be broken by contact with interior projections. It is also difficult to adapt spring pressure to a conduit trolley, so as to prevent its being broken or displaced by the strain caused by obstructions on the track, while the spring pressure and leverage of the overhead trolley easily adjust it to the passage of such obstructions.

The comparative expense of the conduit system is also a constant obstacle to its adoption ; electric companies being slow to adopt it so long as there is any hope of inducing municipal governments to permit the construction of the cheaper overhead system. The cost of construction, with a central conduit, is somewhat less than that of the cable road. The relative stability of the track and conduit, and the strain of the street traffic, in addition to that of the cars, must be provided for in the same manner; but the conduit may be much shallower, no space being required for pulleys, as in the cable conduit, while the cost of curve construction is only about one third that of the cable road curves, which must be so constructed as to sustain the great strain of the cable and prevent its displacement; this strain adding materially to the consumption of power also, and hence to the operating expenses, while, in the electric conduit there is no extra strain to be provided for. The electric conduit has also the advantage of being easily kept clean by brooms and scrapers attached to the cars, being free from the obstructions found in the cable conduit.

The Siemens-Halske Conduit Railway.—The first practical electric conduit railway was constructed by the Siemens-Halske Electric Co. in Budapest, Hungary in 1890; and during the first three years of its existence, it has been largely extended, and has now, May 1893, 40 miles of road in successful operation, on which are running 76 motor cars, no trailers being employed.

The method of construction is illustrated by Fig. 87. The conduit, which is under one side of the track, is $13\frac{2}{10}$ inches deep and $11\frac{2}{10}$ inches wide, and is inclosed with concrete, in which are imbedded transverse iron yokes, $3\frac{1}{10}$ feet apart, to which are bolted two rails, embracing a slot $1\frac{3}{10}$ inches wide, adapted to a flange on the car wheel. The other rail has a groove of the same width, adapted to a similar flange, and is connected with the inner slot rail by iron rods.

The electric circuit is constructed with two light angular shaped iron conductors, one for the positive current and the other for the negative, which are attached to the yokes by insulating supports on opposite sides of the conduit, at the proper hight to be above accumulation of mud and water. Sliding metal contacts, fitted to the angular grooves in these conductors, are connected with opposite poles of the motor circuit by insulated copper conductors inclosed in a shank termed the "ship," which travels in the slot.

The success of this road seems to be largely due to this

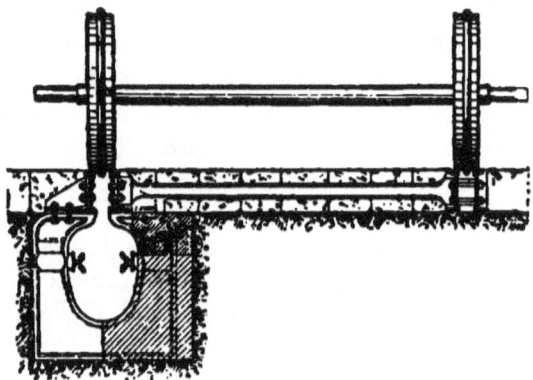

FIG. 87.

method of construction. The large cross-section of the iron conductors makes them equal in conductivity to the comparatively small copper conductors usually employed on electric roads ; while the deep angular groove, with its broad surface, prevents interruption of the electric connection by displacement or breakage of the sliding contact, which is maintained in its place by lateral spring pressure, and hence easily yields to a vertical strain, by which an underrunning trolley would be displaced or broken. The grooved conductors are unobstructed by the projecting clamps, which momentarily break the contact and cause sparking in the ordinary trolley wire construction.

The wide slot is not an essential feature of the construction, but merely a matter of convenience to adapt the track to the flanged wheel and grooved rail in general use in Europe, where narrow carriage wheels, liable to drop into such a slot, are not permitted; the narrow slot required in conduit roads in the United States being equally as well adapted to the electrical and mechanical conditions, and affording better protection for the conduit against the mud of our streets. Our step rail is also better adapted to our muddy streets than the grooved rail, which, it has been found, would require more than double the power for car propulsion, since it could not be kept properly clean.

The climatic conditions are about the same as in most parts of the United States, and the crowding of the cars, usually prohibited in Europe, is permitted in Budapest; so that the operation of the road may be regarded as a fair test of what may be accomplished with a similar system here.

It is claimed that this road is operated with less than half the power of any overhead line, and that while the fare is only $2\frac{1}{2}$ cents, about half that paid in American cities, the net earnings are 55 per cent more than the operating expenses; the cost of coal being about double its average cost in this country, and the cost of labor about half. The cars are constructed in about the same manner as our cars, and the motors and other apparatus are the same as those employed in the overhead system. Gearless motors have recently been put on some of the cars to test their practical efficiency.

The Love Conduit Railway.—A conduit railway invented by J. C. Love was put in operation by the North Division Railway Co. of Chicago in 1892, on a loop line $1\frac{3}{8}$ miles in length, which connects with the company's cable line.

The general construction is similar to that of the cable road. Transverse iron yokes imbedded in concrete, 4 feet apart, support the track rails and inclose a central cast-iron conduit, 9 inches wide by 15 deep, above which is a slot $\frac{3}{4}$ of an inch wide, inclosed between two steel rails, each 5 inches wide, and $\frac{1}{2}$ an inch thick, and shaped like an inverted **L**, having a flange $3\frac{5}{8}$ inches deep on its inner edge, and $1\frac{1}{4}$ inches deep on its outer edge, between which is a projection on the yoke to which it is clamped, and on which it can be moved sufficiently to maintain the slot at its proper width. The upper surfaces of these slot rails are studded with flat projections, each about an inch square, and $\frac{1}{4}$ of an inch high, which serve the double purpose of affording a hold for the clamps and preventing horses from slipping.

The electric conductors are placed directly under these rails, just inside the lower edge of the wide flange, on each side of the slot, and consist of two No. ooo copper wires, relatively positive and negative, attached to clamps, as in the overhead system, which are supported on each alternate yoke by insulating blocks composed of mica and shellac, sustained on rods projecting from the yokes under the slot rails. The conductors are thus amply protected from rain, snow and mud in spaces inclosed by the slot rails both above and at the sides, in the upper part of the conduit above any ordinary accumulations of mud and water.

An underrunning trolley is employed with one wheel on each side, maintained in contact by vertical spring pressure. It is constructed with a steel shank, $18\frac{1}{2}$ inches long, $4\frac{1}{2}$ wide, and $\frac{1}{2}$ an inch thick, which travels edgeways in the slot and supports a cross-bar at its lower extremity to which are hinged two short arms, on the upper ends of which the trolley wheels are mounted. These arms are supported by two spiral springs at an angle of about 30° with the shank; the springs being attached in front to the arms above and the cross-bar below, and curving backward in the same direction as the arms slant, so that when the arms are pressed downward the springs are extended. The wheels are thus given a vertical range of 4 inches, and are attached to the arms by hinge joints, which permit lateral movement at a slight angle; lateral movement of the arms is also provided for at the lower joints, these two movements being required in rounding curves. The trolley is supported on the car trucks by a horizontal bar about 2 feet long, to the center of which the shank is attached. This bar extends lengthways of the car, and is supported at its front end by a hinge, and at its rear end by a cross bar, on which it can slide to the right or left, with the trolley, in rounding curves. Each trolley arm is supported on an insulating strip of vulcanized fibre, and electric connection with the motors is made by insulated copper wires inclosed in two channels in the shank, two wires in each channel; the shank being made of two plates, grooved to form these channels and riveted together.

The cars were propelled by Westinghouse single reduction motors, one motor to each car, geared to the rear axle. Each motor car took one trailer car, and the cars ran continuously round the loop in the same direction, 6 minutes apart, passing four curves on rectangular crossings. There are no grades, and the number of passengers carried was comparatively small, so that while the generator could supply a current of 500 volts, the average current required did not exceed 100 volts.

The operation of the road, since discontinued, was satisfactory. Nothing occurred to interfere with the running of the cars; and it stood the severe test of flooding of the conduit at one spot by the stoppage of a sewer, during which the water came above the wires; but the short circuit thus produced did not cause sufficient loss of current to make any perceptible difference in the operation of the motors, which may be attributed to the low E. M. F. and large current.

This conduit has also been adopted on the Rock Creek railway, Washington, D. C., for that part of its line included within the city limits; and it has been in successful, practical operation since its completion, March 1, 1893.

The conduit is wider than the one in Chicago, so that it is impossible for the trolley, in case of accidental displacement, to strike the sides, and its lower part is free from all interior projections, so as to be easily cleaned.

A very important improvement has also been made in the method of clamping the line wire. The wire is made with a deep groove on each side, into which the jaws of the clamps fit, leaving more than half the wire below them; so that the trolley wheel is in continuous contact with the wire, and jumping over the clamps, causing sparking and liability to displacement of the trolley in rapid running, as in the old system, is thus entirely avoided. The clamps are confined by screw bolts with nuts, and hence are easily tightened, or removed for repairs.

Closed Conduits.—Various methods have been proposed and patented for maintaining electric connection between the line wire and motor either in a closed conduit, or in a closed tube within a conduit, where the wire can be properly insulated and fully protected from liability of contact with the mud and water of the street. But the only one of these which has come into practical use is the conduit system invented by Malone Wheless, described as follows.

The Wheless Conduit Railway.—This system, after a successful experimental test of three years, has been adopted on that part of the Washington and Arlington railway within the city of Washington, and extending three quarters of a mile on the Virginia side to its junction with the part constructed on the overhead system.

The conduit is 17 inches wide and 16 inches deep, and is constructed in the center of the track, with heavy transverse iron yokes supporting two slot rails, between which is a slot $\frac{3}{4}$ of an inch wide. The line wire is supported directly under the slot on vertical clamps imbedded in heavy, hooded insulators, supported on side brackets attached to the yokes. It is composed of No. 4 hard drawn copper wire, and is constructed in sections about 200 feet long, which are insulated from each other by blocks of slate, about five inches long, fitted in slots, and making a continuous coupling between the ends of the wire for the support of the trolley wheels.

The line current is carried in a lead-covered cable of No. 0000 gauge, inclosed in a wooden casing sunk in the earth about six inches inside of one side of the track. From this cable a lead-covered copper feeder extends to each section of the line wire. This feeder passes through an iron switch-box, about 12 inches square and 14 inches deep, which has a water-tight cover, and is sunk in the ground at one end of each section, between the cable and the line wire. This box contains an electromagnet having three poles arranged in a row, and an armature to which is attached a straight bar switch, with a carbon faced contact at each end. This switch closes an opening in the feeder, when the attached armature is attracted by the electromagnet, admitting current from the cable to the line wire section.

The coils inclosing the two outer poles of this magnet are connected together, and their opposite terminals attached to the track rails, completing the electric circuit of a two-cell storage battery carried under the car. The coil inclosing the central pole is connected with the switch terminals, and hence is included in the main circuit when the switch is closed. This circuit passes through a cut-out carried under the car, composed of a switch controlled by an electromagnet, by which the storage battery circuit is opened as soon as the main circuit is closed ; both circuits being connected with the controllers on each end of the car.

When either controller is turned for the admission of

current, it first closes the storage battery circuit, by which the feeder switch is closed and the main current admitted to the line section. This current passing through the cut-out on the car, opens the storage battery circuit, while the feeder switch is kept closed by the central magnet pole, through whose coil the main current now flows. When the car is stopped by the opening of the main circuit, at the controller, the feeder switch is released and current ceases to flow to the line section.

Hence it will be seen that the storage battery circuit, which is entirely independent of the main circuit, is closed only during the brief moment required for the admission of the main current to the line section, and is then immediately reopened by this current; so that the use of the battery current being very limited, a charge lasts several weeks; only one cell being employed at a time, while the other is meanwhile being charged by connection with the line circuit.

Each section of the line is employing current only while the car is passing over it; the feeder switch being opened by the cessation of the current, when the car has passed to the next section, in the same manner as by the stoppage of the car; and this cessation of the main current closing the cut-out switch in the storage battery circuit, current is admitted to this section in a similar manner; this operation being repeated instantaneously at each succeeding section. The return circuit to the central station is made by the rails in the usual manner.

The trolley has two wheels, one traveling in advance of the other; so that in passing from one line section to another, the front wheel makes contact with the section in front before the hind wheel leaves the rear section, hence there is no interruption in the supply of current. The wheels are kept in contact with the wire by spiral springs which press them down and allow a vertical play of about

1½ inches to compensate for inequalities in the line. They are insulated from the traveling shank and connected with the motor by insulated conductors; the current being collected by carbon bearings which make contact with their sides. The trolley is supported on the car truck on two transverse steel guides, by which sufficient lateral play is given for adaptation to curves, some of which have only 48 feet radius.

The object of carbon contacts, both for the wheels and feeder switches, is to prevent sparking and heating, at the contacts, by the passage of the current through an imperfect contact, by which metal switch contacts would be liable to become welded. The carbons on the switch contacts are not subject to wear, and therefore do not require renewal unless injured, and have given very satisfactory practical results.

As there is no current in the line wire except in the sections over which cars are passing, and only in that part of each section between the car and feeder switch, electric leakage on the line is reduced to the minimum.

Each car is equipped with two 20 horse-power motors of the latest Edison design, and the usual 500 volt current is employed. The system has been found to operate successfully under the most difficult conditions of track construction, such as steep grades combined with sharp curves.

Elevated and Underground Electric Railways.—The application of the electric system to an elevated or an underground railway is much easier and simpler than to a surface railway, since the electric conductors cannot interfere with surface travel, can be easily insulated, and hence can be placed in any position, under or over the cars, and be constructed in any manner found most convenient, safe, and economical. Electric traction must therefore eventually supersede steam or cable traction on all such railways, being more economical than either, and free from the annoying smoke, steam, drip, and noise of the former; and is already coming into practical use on railways of this construction both in America and Europe.

The Liverpool Elevated Electric Railway.—This railway extends along the Liverpool docks over a horse railway employed for the transportation of freight, and had six miles of double track completed at the time it was put in operation, March 1893. It is constructed with a water-tight floor made of plate iron, laid on transverse plate girders, supported on channeled iron columns, 50 feet apart. The track is laid on longitudinal sleepers resting on the floor, and the electric conductor consists of a central steel rail, $3\frac{1}{4}$ inches wide, shown in Fig. 88, having a broad groove

FIG. 88.

underneath, with flanges on each side, fitted to porcelain insulators covered with sheet lead and supported on wooden cross-ties.

This rail is about $\frac{7}{8}$ of an inch higher than the track rails, and at switch crossings is therefore elevated that much above the cross tracks, and is divided and the ends bent in opposite directions for a distance of about a foot, parallel with the cross tracks, leaving ample clearance space both at the sides of the cross rails and above them; electric connection being made underneath between the separated ends. This connection is made, both here and between the regular sections of the conductor, by stout copper bands bolted to the flanges on each side.

There are two of these conducting rails, one on each track, which are electrically connected together at the terminals of the road, so as to complete the circuit; the current flowing in opposite directions in each.

The current collector consists of a sliding cast-iron shoe, curving upward at each end as shown in Fig. 88, and hinged loosely to an insulated iron support under the truck. It is much wider than the conducting rail, so as to maintain contact on the bent ends described, at diagonal crossings.

Each train consists of two cars, each 45 feet long, and equipped with a motor at its outer end, so that there is a motor at each end of the train, but only the rear motor is employed for propulsion; the motor man operating it from the front end of the train, and changing his position to the opposite end at each terminal of the road, and reversing the direction of the train. Hence one of the motors is at rest, on each alternate trip, while the other is in operation.

These motors are series wound and gearless, the armature being mounted on the car axle. The maximum armature speed is 260 revolutions per minute, calculated for a maximum train speed of 26 miles an hour; 600 revolutions for each car mile. The actual running speed is 12 miles an hour, including a half minute stop at each of the 15 stations; and the trains are run at intervals of three minutes.

Each loaded train weighs 40 tons, and each car seats 57 passengers, and is lighted by six incandescent lamps supplied with current from the same source as the motors.

The power station is located near the center of the line, and is equipped with six boilers by which steam is supplied to four engines, each of 400 horse-power, which operate four Elwell-Parker, shunt wound dynamos, each furnishing a current of 475 amperes, at an E. M. F. of 500 volts.

Each dynamo is connected through an ammeter with a double-pole, automatic, magnetic cut-out, which is also employed as a main switch; and all the dynamos are connected in parallel with bus bars from which current is taken by underground armored cables to the line conductors; a large automatic cut-out being inserted, which carries the entire current. The E. M. F. of each dynamo is regulated by a rheostat in its field circuit.

The cars are furnished with hand brakes, and Westinghouse air brakes, the latter operated by compressed air stored in receivers carried under the cars, and charged at the end of each round trip.

The City and South London Underground Electric Railway.—This road, which was put in operation in December 1890, extends from King William's statue on King William Street in the city, under the Thames to Stockwell, and is constructed with two parallel iron tunnels, each ten feet in interior diameter, and about 60 feet below the surface.

The line conductor is a steel rail of high conductivity, channeled underneath, and supported on glass insulators in the bottom of the tunnel, between the track rails. It is divided into sections for convenience in testing and making repairs, and these sections are connected by fish plates and copper strips. Its insulation is so perfect that the entire electric leakage does not exceed one ampere, the total loss being less than one horse-power, which is a fraction of one

ELECTRIC RAILWAYS AND RAILWAY MOTORS. 199

per cent of the total power required for the full working capacity of the line.

The current is collected by sliding shoes of iron or steel, from which it passes through an ammeter to a regulating switch, then to a reversing switch, and thence to the motors, returning to the dynamo by the rails, which complete the circuit.

Each train consists of an electric locomotive and three passenger cars. The locomotive, shown in section in Fig. 89, carries two gearless motors whose armatures are

FIG 89.

mounted on the car axles, the armature shaft constituting the axle, and each operating independently of the other. Each locomotive weighs ten tons and has a maximum effective energy of 100 horse-power, and is calculated for a speed of 26 miles an hour. Each is equipped with a Westinghouse air brake, and also a hand brake, and carries a reservoir charged at the depot, at each return trip, with sufficient compressed air for the 40 stoppages made during the round trip.

The entire weight of the train when loaded is 42 tons.

Each car is 32 feet long, seats 34 passengers, and is lighted by incandescent lamps by the same current which supplies the motors. Ten trains are run, following each other at intervals of three minutes.

The power station is located at Stockton, the suburban terminus of the road, and is equipped with three Fowler, vertical, compound engines of 375 horse-power each, which operate, independently, three Edison-Hopkinson dynamos. These dynamos can be operated either as compound or shunt machines. They have bar armatures, each weighing two tons, and bipolar field-magnets, each with its yoke weighing 11 tons, and wound with nearly $1\frac{1}{2}$ tons of copper wire; the entire weight of the dynamo, including its other parts, being 17 tons. Each can generate a current of 475 amperes, at an E. M. F. of 500 volts. They have an electric efficiency of over 96%, and the electric energy available for running the cars is about 75% of the indicated horse-power of the engine. The current is conveyed from the dynamos to a switch-board, equipped with the usual measuring instruments and other apparatus, from which it is distributed by copper cables to the various parts of the line.

Access from the depot, at Stockwell, to the tunnels, is by a curved tunnel, having a grade of 1 foot to $3\frac{1}{2}$, through which the trains are lowered or hoisted, as required, by a rope operated by a stationary engine.

There are 20 neatly constructed stations, each furnished with a hydraulic passenger elevator operated by water supplied from the depot at a pressure of 1200 pounds to the square inch.

Storage Battery Traction.—Instead of conveying electric energy from the power station to the cars by conductors, it may be carried on the cars in batteries in which it has previously been stored. In this way the construction and maintenance of a specially designed electric road may be entirely avoided, and electricity be employed to run the cars on any ordinary horse-car line.

A storage battery is not a generator of electricity, but an accumulator which can be electrically charged by a generator so as to furnish a current of any required strength in proportion to the number of cells and rate of discharge, until the charge is exhausted, when it can be recharged and discharged as before, and this process continued indefinitely till the battery is worn out.

The Faure cell represents the most common method of construction. It has two sets of plates, one positive and the other negative, made with lead grids filled with a paste composed of lead oxide and sulphuric acid; the red oxide known as minium being used for the positive plates, and the yellow oxide known as litharge for the negative. These plates are usually about 9 to 10 inches square and $\frac{1}{4}$ of an inch thick, and are placed vertically in the cells, about $\frac{1}{8}$ of an inch apart, positive and negative alternating, each set being connected together above by a transverse lead bar, and the two sets insulated from each other by rows of hard-rubber forks, and bound together by rubber bands. The two outside plates are always negatives, and are equal to but one plate, their outside surfaces being neutral.

The plates are prepared for use by the electro-chemical reduction of the minium in the positives to lead dioxide, and the litharge in the negatives to spongy lead, with a dynamo current, while immersed in water acidulated with 36% of sulphuric acid; a process which requires about 24 hours for the positives and six days for the negatives. Each subsequent charging after discharge, when in use, produces the same chemical changes, and is accomplished in the same manner, the same kind of fluid being employed, but requires only a few hours. The discharge is produced by the external electric connection of the two sets of plates, which may be made through a motor or any piece of apparatus to be electrically operated, and results in the chemical reduction of the lead dioxide and spongy lead to lead sulphate on both plates, with the absorption of sulphuric acid, and the evolution of electric energy equal to about 80% of that employed in charging.

The 23 plate cell, with a closely covered hard-rubber vessel, is preferred for car batteries. A battery of about 100 cells is required to propel an ordinary street car. Its weight is about 4000 pounds, its E. M. F., when connected in series, 200 volts, and its storage capacity 150 ampere hours. It can be charged in about 8 hours by a dynamo current of 10 to 25 amperes, and will propel a car for about 6 hours, with a current of 15 to 30 amperes; minimum current being required on the level, and maximum in starting and ascending grades; 60 amperes being sometimes required to propel a heavily loaded car on an ascending grade.

Two batteries are required for each car, one being charged at the power station while the other is in use on the cars: but as the time required for charging is about one fourth longer than for discharging, it is evident that if the batteries were always fully charged and discharged, extra batteries would be required to keep the cars running; hence it is found better to limit the time of discharge to about 3 hours, during which a run of about 25 miles can be made at the ordinary speed permitted within city limits; and if the fresh battery is not ready, the time can be extended without injury to the battery; whereas if the battery were employed longer than the normal time of discharge, it would be liable to serious damage; an exhaustive discharge being injurious while an overcharge does no harm.

The battery is placed under the seats, in sections mounted on trays furnished with casters, and can be easily and rapidly moved on or off through openings in the end of the car; three minutes being the average time required to exchange a discharged battery for a charged one. The electric circuit and its equipment is the same as on other electric cars, but by arranging the battery in four sections, which can be connected in series or in parallel, the E. M. F. and relative current volume can be varied inversely as required. The four sections being connected in parallel when starting, a current of 50 volts and relatively large volume is obtained; by another move of the switch, the battery is connected in two parallel circuits, and a current of 100 volts and half the volume obtained; and by a third move, when the car is fully under way, the four sections are connected in series, and a current of 200 volts and one fourth the volume obtained. This arrangement does not interfere with the control of the current through a rheostat in the usual way.

This system is still in the experimental stage, and has made comparatively small progress. That it is much more expensive than the overhead system seems to be clearly demonstrated, and its principal chance of success is in competition with the conduit system. If it shall be found more practical or economical than the latter, it may be able to compete successfully with this, or other methods of traction, where the overhead system is prohibited. Its power station and motors cost just the same for construction and maintenance as either of the other electric systems, and its only economical advantage is in line construction, and its adaptation to use on horse-car lines already constructed, as has been noted. The frequent renewal of the batteries is the largest item of expense; the average durability of the positive plates not exceeding one year while that of the negatives is much longer. The great battery weight to be carried, in addition to the weight of the motors, adds materially to the consumption of power and wear of the rails, and hence to the current expenses. The lead batteries described, which are the kind commonly employed, are not adapted to use on roads where the grades exceed 5 feet in the 100; the electric strain on roads with steeper grades being too great for them.

Various improvements are being made in storage batteries, one of the most recent being an improvement of the original Plant battery, the first storage battery invented. In the improved construction, grids made with transverse folds of corrugated, flexible, lead ribbon inclosed in lead frames, are employed instead of the stiff grids with rectangular holes used in the Faure battery. And the grids, instead of being filled with paste subsequently reduced to lead dioxide and spongy lead, as described, have these materials rapidly formed in them by immersion for a few hours in a fluid of special composition, producing plates of more homogeneous composition than the paste-filled plates. This construction, it is claimed, materially increases the battery's efficiency and durability.

The fact that nearly every attempt to run cars by storage batteries has been abandoned after trial, shows that the system is still in a very imperfect state, though experimental work is still being continued which may eventually lead to practical results.

Electric Haulage in Mines.—For the transportation of coal and ore in mines railways are required on which cars may be propelled either by animal power, rope haulage with stationary power, steam locomotives, or electric locomotives. The superior advantages of the last method over animal or rope haulage in convenience and economy has been fully demonstrated, and its superiority to the steam locomotive in furnishing power without vitiation of the air by combustion, smoke, steam and gas, is sufficiently apparent; the steam locomotive being confined to the return air passages, which convey the vitiated air out of the mine, where its use is least objectionable, while the electric locomotive can be employed in any part of the mine, and is entirely free from every element which could interfere with the comfort of the miner, injure health, or menace life or property; electric current being supplied at the comparatively low pressure of 220 volts, which insures perfect safety.

While the reciprocating motion of steam locomotives limits them to grades not exceeding 4 feet in the 100, the rotary motion of electric locomotives enables them to ascend grades of 12 or even 14 feet in the 100, such as are common in mines; and where the grade is in excess of this, in ascending from one level to another, rope haulage with hoists operated by stationary motors is employed to move the cars up the incline.

This system is adapted not only to the movement of heavy trains on the main trunk lines, but may also be employed in making up the trains; the cars in long side entries being hauled by locomotives equipped with small motors more economically than by animal power, which is the usual method.

These roads are usually of narrow gauge, ranging from 18 to 36 inches in width, and traverse low passages; and require low narrow locomotives, of massive structure, completely iron clad for protection against dripping water, flooded tracks, and accidental obstructions, and adapted to draw heavy trains at a speed not exceeding 6 to 10 miles an hour. A single motor is usually employed, mounted between the axles, and connected with both by double reduction gearing. The motors range in energy from 10 to 100 horse-power, and the locomotives in weight, from 3500 pounds to 21000 or more. The type known as "Terrapin Back," as made by the Thomson-Van Depoele Electric Mining Co., has a square frame with flat sides from which project cases inclosing the wheels; it is high at the center over the motor, and curves downward at each end, where a space, inclosed at the sides, is provided for the motor man, who sits within reach of the controlling apparatus and brake. The motor is series wound, has a Gramme ring armature, a field with two pole-pieces and consequent poles, and carbon brushes set radially, at which there is no sparking.

The line is usually constructed with a trolley wire above, an underrunning trolley, and a return circuit by the rails; but as the track is generally cheaply constructed, not having the permanency of a street railway, and hence liable to interruption of electric continuity, a supplementary ground wire connected with the rails is sometimes employed, or two overhead wires, one for the positive and the other for the negative current, and an overrunning double trolley connecting both with the motor. The wires are sustained on insulating supports attached to the roof or walls of the passage ways, and the usual switches and crossings provided.

The general construction, equipment and operation of these railways being similar in other respects to those of electric street railways, further details are unnecessary. The different electric companies have constructed and equipped such lines in various large mines in different parts of the country, which are now in successful operation; and a similar system is also in use in many of the European mines.

Electric Haulage in Mills and Factories.—The transportation of raw material, manufactured articles, and fuel from one part to another of large manufacturing establishments requires tramways several hundred or thousand feet in length, which traverse the yards and buildings, usually with numerous sharp curves and often with steep grades. These roads vary in gauge, the narrow 36 inch, or the standard $56\frac{1}{2}$ inch gauge being the most common, and are equipped with tram cars adapted to the various kinds of material to be carried. Animal power being expensive and inadequate to the requirements of a large modern factory, either steam cars, electric cars, or cable traction with stationary power is required. Here, as in the mine, the superior advantages of electric traction over steam or cable traction are shown in economy, safety, and adaptability to existing conditions. The use of the steam car is limited to the yards, and even there its noise, smoke, and escaping steam are highly objectionable, and its flying sparks dangerous; while the electric car, noiseless, clean, and safe, can follow the track through buildings and yards, and up steep grades, impossible of ascent by the steam car.

The lines are constructed in the usual manner, with an overhead trolley wire and return circuit by the rails; the wires being suspended at such hights as may be required by street crossings or low passage ways through buildings, the range being from 6 to 18 feet: and current is supplied at a pressure of 220 volts by dynamos operated by the power which runs the factory.

The cars are equipped with either one or two motors, ranging in energy from 3 to 30 horse-power. The motor car usually carries load and also draws other cars. It is also sometimes equipped with a hoist which can be connected with the motor by special gearing, and is employed to haul small loaded cars on side tracks up to the main track, where they are transferred to trucks and hauled to any required point; the side tracks being at right angles to the main track, and elevated so as to be on a level with the platforms of the trucks on the main track; this track being of a special gauge adapted to cars wide enough to receive the small cars transversely.

This system is now in successful operation in many of the large cotton and woollen mills of New England, and in iron, steel, glass, and machine works in various localities. It is also employed for switching freight cars and making up trains; also for surface haulage in connection with mining work, for the transportation of coal or ore.

CHAPTER VI.

CENTRAL STATION CONSTRUCTION AND EQUIPMENT.

Development of the Central Station.—The electric central station, as now constructed and equipped, is the result of the gradual development of the electric system through the several stages of arc lighting, incandescent lighting, power applied by the stationary motor, and power applied by the railway motor. Previous to 1888, central stations were constructed for electric lighting alone, the demand for current to operate electric motors not being sufficient to justify the investment of capital for its supply. But the success of the motor as a means of applying power soon increased this demand, and electric companies recognized the importance of meeting it by the enlargement of their central station equipment, so that capital invested in buildings and apparatus, which was comparatively idle during the day, when but few lamps were in use, could be profitably employed in operating motors. The accomplishment of this was comparatively easy, since the demand on the engines and generators for this class of work was not so materially different from that for electric lighting as to require generators of special construction; the current being comparatively steady, without excessive fluctuations.

The generation of electric energy for both power and light in the same station has aided very materially in bringing the stationary motor into use, as the demand for motors is not yet sufficient, even in large cities, to justify the maintenance of a separate station and a separate system of underground distribution. But the conditions of electric railway work are so different from those of stationary motor work, that it has not been deemed practicable to employ the same station for this purpose also, though in many instances, the important advantages in economy, and the control of valuable franchises render it probable that such a combination will be effected. The central point of distribution for an electric railway is often quite remote from that of a profitable electric lighting and stationary motor district, and on land of much less value. The electric energy required for the running of a given number of cars can be easily calculated in advance, while that required for stationary motors and lighting is an uncertain quantity dependent on public patronage. The electric strain on railway generators, like that on railway motors, is liable to excessive fluctuations, and must be provided for in a similar manner, and the steam engines must also be proportioned to the demand of the generators upon them ; so that generators and engines adapted to electric lighting and stationary motor work would be quite unsuitable for railway work.

The construction and equipment of the central station, and the adaptation of its apparatus to the specific purposes mentioned will be best understood from the following practical examples.

Chicago Edison Central Station No. 1.—The principal station of the Chicago Edison Company, known as No. 1, is located at 139 and 141 Adams Street, near the center of the principal business district of the city. It has two main stories, which contain the machinery and apparatus, and a limited space in front for the offices. The engine room, which is on the ground floor, is about 46 × 100 feet, the dynamo room above it is the same size, and in the rear of each is a smaller room for the boilers.

There are 6 Armington and Sims engines, 2 of 500 horse-power each, and 4 of 250 horse-power each; 6 McIntosh and Seymour engines of the same relative sizes; and 2 Ideal engines of 250 horse-power each; 14 engines in all, representing 4500 horse-power. Steam is supplied to them by 9 Heine safety boilers, 4 in the lower boiler room and 5 in the upper; and the average daily consumption of coal is about 75 tons.

There are 28 Edison, 110 volt dynamos, 20 of 125 horse-power each, and 8 of 250 horse-power each; making a total of 4500 horse-power. These are placed in the dynamo room, directly over the engines, to which they are belted through openings in the floor; each engine operating 2 dynamos. These dynamos are of the standard Edison shunt wound type, similar in construction and appearance to the standard Edison stationary motor, described in chapter III. They are large massive machines, the larger 8 feet in hight, and the smaller 6 feet.

They are connected together in pairs, in accordance with the Edison three-wire system; the positive pole of one dynamo being connected to one side of the main external circuit, and the negative pole of another dynamo, to the other side; the two remaining poles being each connected to a central main, known as the neutral wire. These mains extend through pipes under the streets, and thence into the buildings, and the lamps and motors are placed on parallel circuits connecting the central main with the exterior mains on each side; the arrangement being similar to that of a double ladder with rounds connecting a central bar with two side bars. If the current consumption is equal on both sides, no current flows through the neutral wire; but if the consumption is greater on one side than on the other, as generally happens, the excess of current required for that side flows through this wire, either in a positive or negative direction, according to the side requiring the supply.

On each side wall of the dynamo room is a switch-board extending nearly the entire length of the room; on which are mounted in parallel rows, the ammeters above, the switches below them, and the rheostats in the lowest row. The principal connections are made with flat copper bars, without insulating covering, which extend across the ceiling and down the boards on each side, connecting with the instruments and with cross-bars; connection with the dynamos being made by vertical wires from the ceiling bars.

The board on the east wall, shown in Fig. 90, is divided into four sections; three embracing the connections for eight dynamos each, and the fourth those for four of the large dynamos. In each section half the dynamos are connected in parallel to the positive side of the three-wire system, and half to the negative, each by one of its armature poles; the connections being made with the neutral omnibus, or "bus," bar, as shown, which is tapped by feeder wires connecting with the neutral wires in the street.

There are 28 ammeters and 28 knife switches on this board, one of each for each dynamo, through which the armature circuits connected with the neutral bus bar pass; the ammeters on the positive side, in each section, indicating the same current as the corresponding ones on the negative side. In each section are two potential indicators, each represented by two little rectangles, one for the positive and the other for the negative side, and the eight are connected in series with two small wires which connect with the mains at a distant point in the district, showing the E. M. F., or electric pressure, at that point, relative to a fixed standard by which it is regulated at the dynamos. Two would indicate the pressure as well as eight, but would not be so convenient for observation on a long board. There are also two voltmeters, indicated by the letters V M in the circles on the right, one for each side of the circuit,

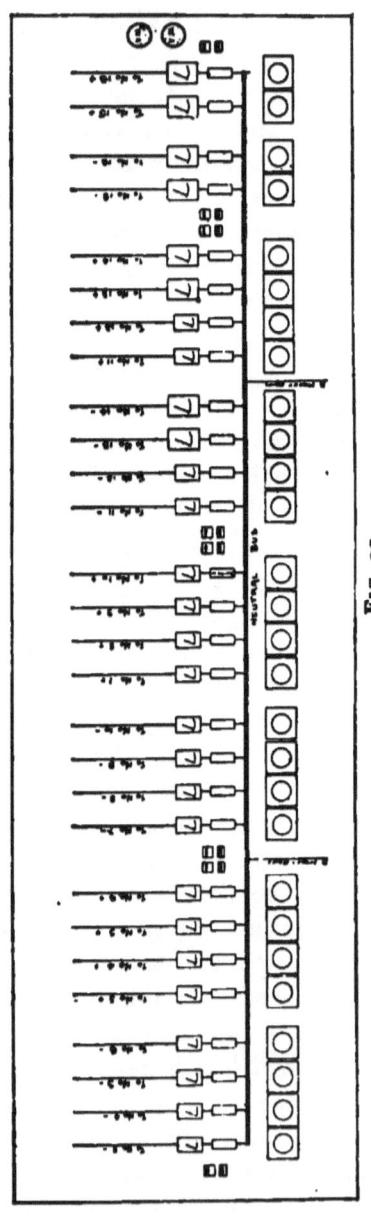

Fig. 90.

by which the E. M. F. between the opposite sides, at the station, is measured once every day. There is also a rheostat for each dynamo, by which its field current is regulated, and which is connected with its field circuit by two small wires.

On the west switch-board, shown in Fig. 91, are mounted two parallel bus bars, each being the full length of the board; and the 14 dynamos on the positive side of the system are connected in parallel to one bar, and the 14 on the negative side to the other, each by the armature pole opposite to that connected with the neutral wire on the east board; only five pairs of these connections being shown in Fig. 91. Each of these bars is tapped by feeder wires, connected with the street mains which extend to different sections of the district; the connections being made through ammeters and switches, as on the east board. Lightning arresters are not required, as the circuit is under ground.

The course of the armature current is as follows:— Starting from the positive poles of the dynamos on the positive side, it flows to the positive bar on the west board; thence through the feeders to the lamps and motors throughout the district; back through the feeders on the negative side to the negative bar; thence to the negative poles of the dynamos on the negative side, and through their armatures to the positive poles; thence to the east board, through its connections, and back to the negative poles of the dynamos on the positive side, and through their armatures to the positive poles, completing the circuit; the excess of current required on either side flowing through the neutral bus bar and feeders in the manner already described.

As the dynamos are all connected with the circuit in parallel, any number of them may be cut out in pairs, by opening their switches, without interference with those remaining

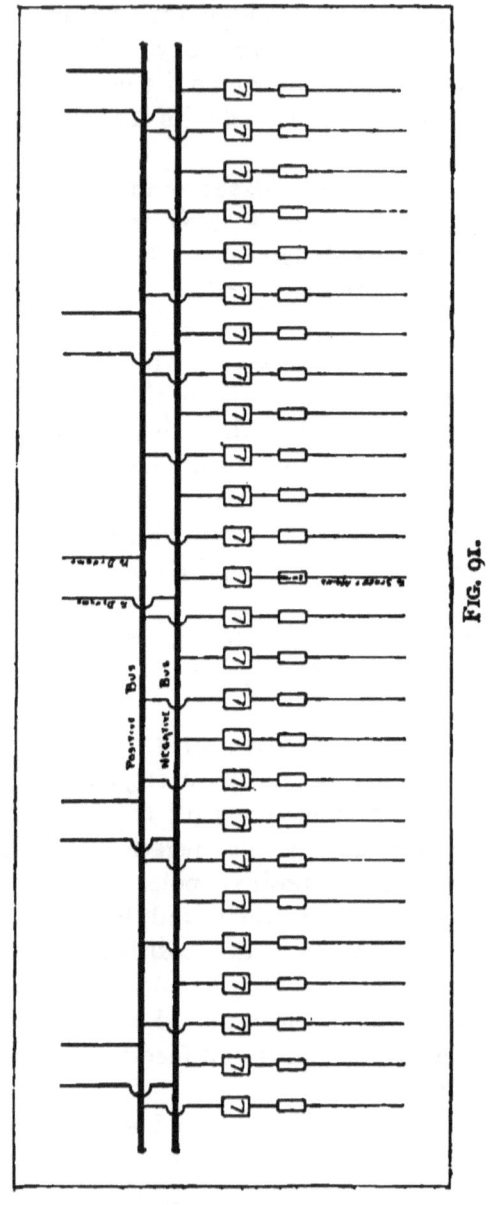

Fig. 91.

in operation, the engines connected with them being also stopped, and the supply of current thus regulated in proportion to the demand. Thus during the day, when current is required only for motors and a comparatively small number of lamps, many of the dynamos and engines remain idle which are employed during the maximum demand for light in the evening.

The district embraces an area of about a square mile, in which 700 arc lamps are supplied with current, 50,000 incandescent lamps, and about 350 motors, besides an indefinite number of small motors, used on lamp circuits, of which no record is kept.

Chicago Edison Central Station No. 5.—This station recently purchased from the Chicago Arc-Light and Power Company is at 250 Washington street, near the center of a district embracing about six square miles of territory; the major part of its business being done in the same territory as that of the station already described. It is a large four story building, occupied both by the machinery and offices. The dynamo room, which is on the second floor, is 54×140 feet, the engine and pump room, below it, is the same size, adjoining which is a smaller room for the boilers.

There are 5 Williams compound, condensing engines; 1 of 250 horse-power, 1 of 600, and 3 of 500 each, 2350 horsepower in all; which are supplied with steam by 10 Manning upright boilers. The daily average consumption of coal is about 35 tons.

There are 58 dynamos, representing 2328 horse-power; 2 Standard, series wound, of 2500 volts and 40 horse-power each, 6 Sperry, series wound, of 1500 volts and 20 horse-power each, and 50 Thomson-Houston, 1 of which is a multipolar, compound wound machine, of 500 volts and 104 horse-power, 3 compound wound, of 500 volts and 68 horse-power each, 3 alternating current, of 1000 volts and 94 horse-power, each excited by a small direct current dynamo of 3 horse-power, 43 series wound, 2 of which are of 1750 volts and 24 horse-power each, and 37 of 2500 volts and 40 horse-power each.

These dynamos have been rated according to their horse-power, for convenience, as this rating applies to all of them; but series dynamos, when employed for arc lighting, as in this case, are usually rated according to the number of 2000 candle-power lights each is capable of feeding, each light representing $\frac{4}{5}$ of a horse-power in electric energy; so that a series dynamo of 40 horse-power would be known technically as a 50 arc-light machine.

Power is transmitted from the five engines to three separate shafts in a room over the dynamo room, two engines being connected to one shaft, two to another, and one to the third, by belts through openings in the floors and ceilings; and from each of these shafts, belts extend down to the dynamos, by which they are operated.

The dynamos are connected, at one end of the room, with an insulating switch-board of slate, shown in Fig. 92, by wire cables, wrapped with insulating material and inclosed in insulating tubes covered externally with brass, which extend along the ceiling in three separate groups. The board is divided into three sections; the section on the left connected with the three alternating dynamos, which are employed for incandescent lighting, the section on the right with the four compound wound dynamos, employed for motor work, and the central section with the 51 series dynamos, employed for arc lighting.

In the left hand section are mounted, in three vertical rows, one for each alternating dynamo, an ammeter, a potential indicator, a double pole knife switch, and two rheostats, one for the field circuit of the exciter, and the other for its armature circuit, whose current traverses the field circuit of the alternater. On the back of the board, in this section, are three transformers, one for each dynamo,

CENTRAL STATIONS. 217

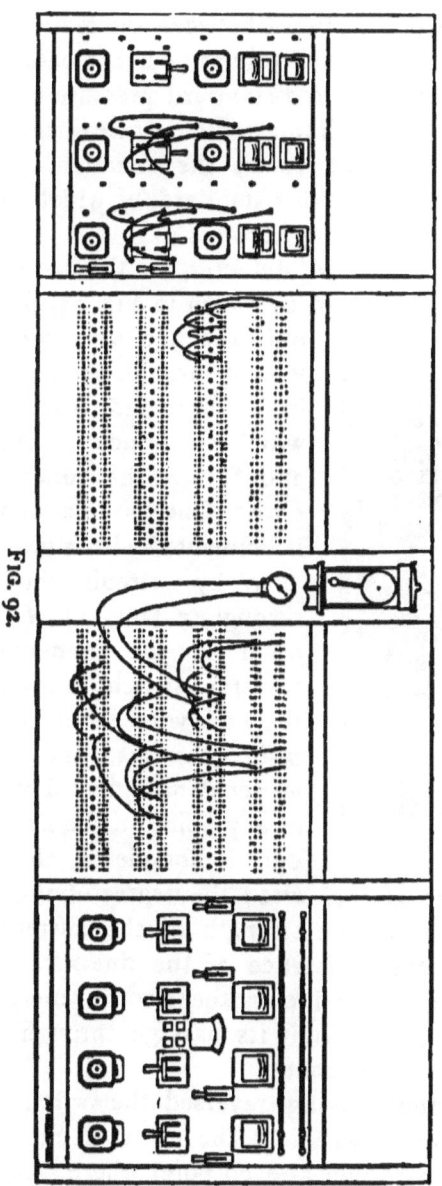

Fig. 92.

connected with a shunt circuit between the mains, through which a small portion of the primary current passes, and is reduced from 1000 volts to a standard pressure of 52 volts in a secondary shunt circuit which passes through the potential indicator, which shows the variation of electric pressure with reference to this standard, by which the pressure can be regulated at the rheostats.

There is also a ground detecter, in this section, constructed like a transformer, as shown in Fig. 93, but having only the fine wire copper coil, wound through a square opening in a laminated iron structure which surrounds it externally, and fulfills the function of a core. One terminal of this coil is connected with a ground wire at *B*, by a circuit which passes through a lamp at *A*, and the other can be connected by a transfer switch, shown under the coil, with either side of the primary circuit; and if there is an accidental ground connection at any point on the line, the circuit is completed, through the detector, with the ground wire; the degree of illumination in the lamp showing the strength of the current in this ground circuit; the resistance of the fine wire and self-induction of the coil reducing the electric energy of the current sufficiently to permit its passage through the lamp without injury to the filament.

FIG. 93.

The main current, having traversed the switch and ammeter on the board, passes into the external circuit, where it traverses a transformer at each point where incandescent lights are required, by which its E. M. F. is reduced from

1000 volts in the primary circuit to 52 volts in a secondary circuit by which the lamps are supplied, and in which they are mounted in parallel between the mains; the strength, or volume, of the current in the secondary circuit being increased in the inverse ratio of this reduction of E. M. F. The two circuits are insulated from each other, and the secondary current is induced in the coarse wire coil of the transformer, by the primary current which traverses its fine wire coil.

In the right hand section, the four dynamos are connected together in parallel; their positive poles being connected with one bus bar and their negative with another; each circuit passing through a knife switch, an ammeter, and a rheostat, and being also connected with a third bus bar by which the current of the four is equalized. There is also a potential indicator, placed on a shunt circuit connected with the two main bars. Each of these bars is tapped by four feeders, making four parallel circuits, each of which is connected with a double knife switch, and extends out into the district, supplying current to motors placed on parallel circuits between the mains.

In the central section, the board is pierced with holes arranged in horizontal and vertical rows, and in each hole is a spring-jack into which a plug connected with a flexible conductor can be thrust. The terminals of the 51-series dynamo circuits are connected at the back of the board with the spring-jacks in the two upper rows, the positive terminals in one row and the negative in the other. In the six rows below these, the spring-jacks are connected with the terminals of the external circuits; three rows of spring-jacks, connected with positive terminals, alternating with three connected with negative. And between the terminals of each external circuit is a transfer switch by which the connections can be changed from one circuit to another; these switches being represented in Fig. 92 by three rows of small circles. The holes are all numbered, on the front of the board, and those in the two upper rows are lettered also in three separate groups, A, B, and C, so that the connections can be properly made.

Each dynamo circuit, in the two upper rows, is connected, on the front of the board, by a pair of flexible conductors terminating in plugs, with an external circuit in two of the lower rows, or with two or more of these circuits connected together in series. The series connection is made by connecting the positive pole of the dynamo circuit to the positive pole of one external circuit, and the negative pole of the dynamo circuit to the negative pole of another external circuit, and then connecting the negative pole of the first external circuit to the positive pole of the second; the two external circuits, thus connected together in series, being practically the same as one circuit having two sections. In like manner any number of external circuits may be connected in series with the same dynamo circuit, provided the whole number of arc-lights in use on them simultaneously does not exceed the electric energy of the dynamo. Thus during the day when but few lights are required, one dynamo may supply several circuits, while, after dark, the connections can be changed to as many dynamos as are required to supply the increased number of lights in use.

There are two small holes on the board adjacent to each large one, in which temporary connections can be made for any purpose without disturbing the main connections. Each external circuit has two pairs of spring-jacks connected with two pairs of the large holes, and by means of the transfer switch, which is placed between them, the circuit can be connected with a dynamo circuit by either pair. When the connections are to be transferred from one dynamo to another, connection is made with the second dynamo in the two vacant holes, and by a turn of the switch, the terminals of the external circuit are transferred from the first dynamo to the second, in an instant, without perceptible interference with the lights on the external circuit.

There are no rheostats, voltmeters, or knife-switches employed in this section, nor any parallel connections ; and but one ammeter, which can be connected by flexible conductors with any circuit whose current is required to be tested ; this connection being made in the small holes, the main circuit being opened during the test by withdrawing a plug from one of the large holes. The lamps on each circuit are connected in series and fed by a ten-ampere current, any variation from which is shown by this test, and can be corrected at the dynamo.

There are 2250 arc lamps of 2000 candle-power each, supplied with current from this station, 2000 incandescent, 16 candle-power lamps, 70 shunt motors, representing, in the aggregate, about 600 horse-power, and 123 series motors, representing about 200 horse-power. The mains are copper cables wrapped with insulating material, and inclosed in lead pipes which extend under the streets in conduits, some of which are composed of an insulating concrete made with asphaltum as its chief material, and pierced with holes for the cables ; and others are wooden pipes, coated with coal tar, which are found to be very durable.

Cicero and Proviso Street Railway Central Station.—The Cicero and Proviso street railway is an electric road constructed on the overhead system, which runs from Garfield Park, Chicago, $5\frac{3}{4}$ miles west to Maywood. It has three branches, two on parallel streets and a third extending to Waldheim Cemetery, the whole embracing about $12\frac{1}{2}$ miles of double track ; on which are run 30 motor cars, 11 trailers, and 12 cars which, by change of equipment, may be run either as motor cars or trailers ; the daily average number in use being about 20. There are 12 cars equipped with Sprague double reduction motors, 12 with Edison single reduction motors, and 6 with Thomson-Houston, water-proof, single reduction motors.

The central station is at Ridgeland, near the center of the line, and comprises the power house, car barn, repair shop, and offices. The power house is a one-story building, about 75 × 120 feet, divided into two rooms, a large one, about 75 × 80 feet, for the engines, dynamos and switch-board, and a smaller one, about 75 × 40 feet, for the boilers.

There are three steam engines, an Armington and Sims high speed, of 200 horse-power, an Eclipse Corliss, of 250 horse-power, and a Bullock Corliss of 600 horse-power. These are supplied with steam by six horizontal tubular boilers, with Rooney stoker attachments. The daily average consumption of coal is about $9\frac{1}{2}$ tons. Each of the smaller engines operates two Edison No. 32, compound wound dynamos of 134 horse-power each, and the larger one, two No. 80 dynamos of the same construction, of 268 horse-power each; all connected with the engines by belts, and furnishing a current of 500 volts E. M. F.

The dynamos are connected by overhead wires with a marble switch-board, shown in Fig. 94, on the lower part of which is a rheostat E, for each, separately connected with its field circuit, by which its field current is regulated. The board is divided into two sections, an operating section, X, on the left, and a distributing section, Y, on the right. The operating section is equipped with a circuit-breaker, A, an ammeter, B, a fuse-block, C, a knife switch, D, and a lightning arrester, for each dynamo, and two voltmeters which may be connected with the dynamo circuits by flexible conductors, as shown, for testing the E. M. F. of the current. And on the back of this section are two bus bars, a positive and a negative, with which all the armature circuits are connected in parallel. The positive bar extends to the center of the distributing section, where it connects with a distributing bar through a 2000 ampere ammeter and

CENTRAL STATIONS.

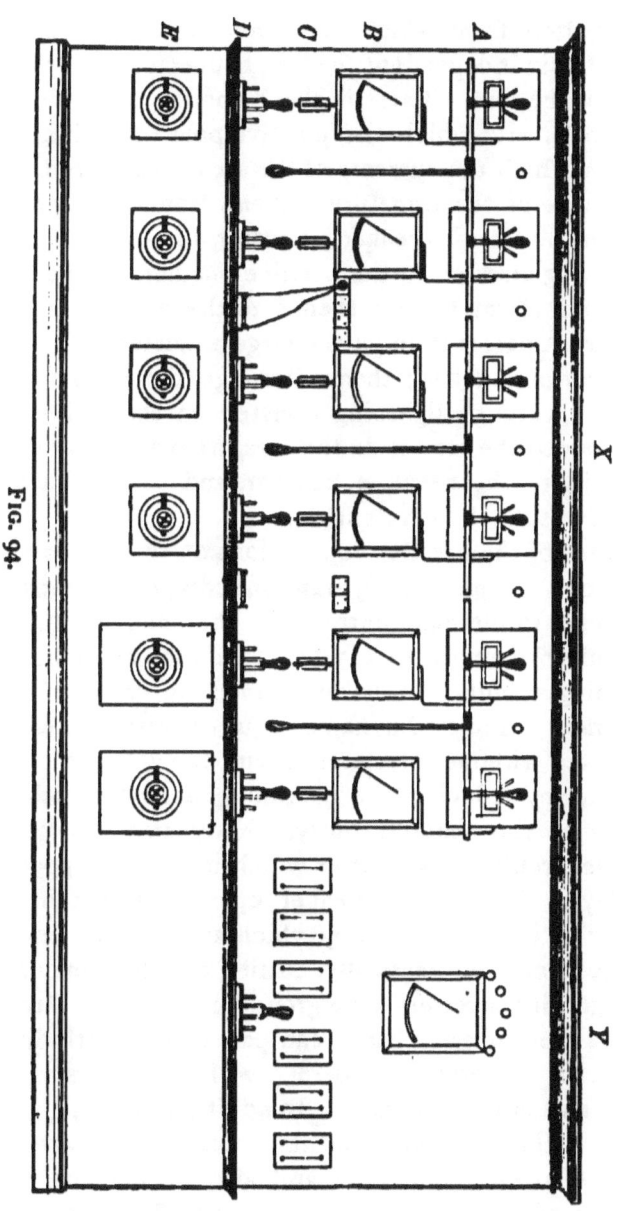

Fig. 94.

a knife switch, from which the connections pass through six fuses to six independent feeders, connected with the six sections of the line. There is also a conducting cable over the dynamos, with which the positive pole of each is connected, by which the currents of the six are equalized.

The course of the armature current, from each dynamo, is through the circuit-breaker, ammeter, fuse-block, switch, and lightning arrester, to the positive bus bar, on the operating section of the board; thence to the distributing section, where it passes through the large ammeter and switch to the distributing bar; thence through the fuses and by the feeders, through lightning arresters on the wall, to the line; thence to the motors on the cars, from which it returns by the rails to the negative bus bar, and thence back to the negative side of each dynamo.

By means of the parallel connections on the bus bars, all the dynamos are made to operate unitedly, each contributing its share to the main current on the line; the generation of current being varied in proportion to the number of cars running simultaneously, by connecting or disconnecting dynamos. And if a dynamo requires repairs, it can be disconnected without interference with the running of the cars, except by reducing the maximum number which can ordinarily be run simultaneously. A dynamo can be cut out by the circuit-breaker, after which its belt is slipped off the pulleys. This circuit-breaker operates a switch controlled by an electromagnet, by which an abnormal current opens the circuit automatically, cutting out the dynamo.

The lightning arresters have ground connections through which the atmospheric charge can pass to the earth direct, by virtue of its great E. M. F., without traversing the dynamos and other apparatus. In addition to the lightning arresters on the board and wall there are lightning arresters on the line, at every half mile, and also on the cars. But this means of protection is not always sufficient against a heavy charge. The board is lighted at night by ten incandescent lamps, shown by the small circles on its upper part, five connected in series on each section.

Waterpower Stations and Long Distance Transmission.—The transmission of electric energy by means of the alternating current renders it practicable to use waterpower for electric lighting, heating, and the operation of motors at points many miles distant from its source. The high E. M. F. of this current, as generated, and its capability of transformation to higher E. M. F. with corresponding reduction of current strength, and to lower E. M. F. with corresponding increase of current strength, makes it a suitable medium for the transmission of power as above; since its high E. M. F. enables it to overcome the electric resistance of a long conductor of comparatively small cross-section, by which power, transformed into electric energy, can be cheaply transmitted to a distant point without great practical loss; and there, by a transformation of E. M. F. to current strength, be applied to practical use.

By establishing a central station at any point where sufficient waterpower can be obtained, and equipping it with alternating current dynamos operated by turbine water-wheels, power which has been unemployed, or limited to local use, can be transformed into electric energy and tributed over a large area, or transmitted to some distant city and furnished often at cheaper rates than it can be generated by steam at the points where it is required.

The distance to which power can be economically transmitted in this manner is still a matter of experiment, involving the relative quantity of E. M. F. and size of line wire which should be employed, each varying inversely as the other; the insulation required by the line and the transformers with reference to this high E. M. F.; and the relative cost of such long transmission as compared with that of short transmission from local steam-power stations.

The Frankfort-Lauffen Experiment.—The most noted experiment of this kind was that made at the International Electric Exposition held at Frankfort-on-the-Main in 1891, where power electrically transmitted from a water-power station at the Neckar-Falls at Lauffen, 112 miles distant, was employed for the operation of electric motors and for electric lighting. The power was transformed into electric energy at Lauffen by a dynamo of 300 horse-power, which generated a three phase, rotary, alternating current, having an average strength of 1490 amperes, and an average E. M. F. of 54 volts. This current traversed a three wire, local, primary circuit, which carried the three successive impulses of the current, produced at each revolution of the armature, and was connected with three transformers, one for each branch of the circuit, by which the E. M. F., at the beginning of the exposition, was raised to 16000 volts, and subsequently to 30000, with corresponding reduction of current strength. The current was transmitted from these transformers to Frankfort by a three wire secondary circuit, composed of bare copper wires, each four millimeters in diameter, less than $\frac{1}{6}$ of an inch, and was received there by three transformers which reduced its E. M. F. to 100 volts, with corresponding increase of current strength, adapting it to practical use. It was employed to illuminate a sign composed of 100 incandescent lamps, each of 16 candle-power, to run several small alternating current motors, and one large one which operated a centrifugal pump, by which water for a large artificial waterfall was raised to a hight of nearly 33 feet; the energy of the Neckar-Falls being thus made to reproduce itself in a waterfall at Frankfort.

The average electric energy generated at Lauffen was about 80500 watts, and that received at Frankfort, about

58000, showing an efficiency of about 72 per cent. The wires were supported on poles by oil insulators, and oil was also employed for insulation in the transformers, with satisfactory results in both cases; showing the value of oil as an insulator where such high E. M. F. is required. There was no perceptible loss of electric energy by rain or fog; the instruments giving the same indications in wet weather as in dry.

Willamette Falls Waterpower Station.—This station is at the falls of the Willamette River in Oregon, 13 miles from Portland, where water-power estimated at 225000 horse-power is obtainable. It was constructed in 1890 for the purpose of utilizing this waterpower for electric lighting at Portland, and was equipped, at first, with two Westinghouse, alternating current dynamos, to which five more have since been added. These are operated by Victor turbine wheels of 300 horse-power each, geared to horizontal shafts, from which belts extend to the dynamos, two of which are operated by each wheel.

Each of these dynamos generates a current of 400 volts; the construction of the armature and insulation of its coils being specially adapted to this high E. M. F. The cores of both field and armature are laminated. The field coils are wound on 12 poles radiating inward from a circular yoke, in the usual manner, and the armature coils on 12 T-shaped core teeth, and insulated from each other by wooden wedges; each coil being wound round the interior shank of the tooth, and covered outside by its projecting flanges, the wedge filling the remaining space between each pair of teeth, so that only the ends of the coils are exposed. The field current is supplied by a small direct current exciter in the usual manner.

The main current is transmitted to Portland by an overhead line consisting of two No. 4 copper wires from each dynamo, supported on poles by glass insulators, and is received at a substation, at an electric pressure of 3300 volts; showing a fall of potential of 700 volts, equal to 17½ per cent in transmission. At this station each circuit is connected with a separate transformer by which the E. M. F. of its current is reduced from 3300 volts in the primary coils to 1100 in the secondary. This transformer is composed of ten sections, or small transformers, connected together in series, as shown in Fig. 95, each of which bears its proportional part of the reduction, so that the E. M. F. in each is

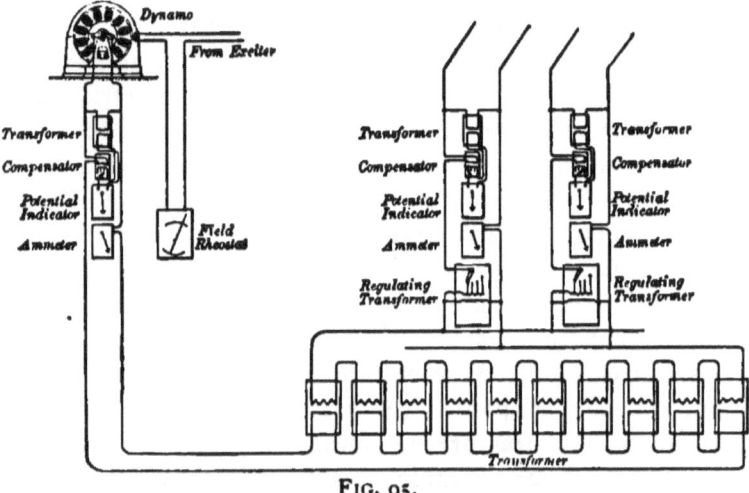

FIG. 95.

330 volts in the primary coil and 110 in the secondary. The difference of potential between each primary and secondary coil being thus reduced to $\frac{1}{10}$ of what it would be in a transformer with two large coils, permits the use of proportionally larger wire, and gives more space for insulation between the coils, for which wood is used in this transformer. A section may be cut out, if required, by short-circuiting its terminals, without affecting the total reduction of E. M. F.; the potential difference in each of the other sections being proportionally increased; a similar result being produced by

the accidental short-circuiting of a coil. The transformer's efficiency is 96 per cent, at full load.

Feeder circuits connected in parallel with the secondary circuit of each transformer, as shown, distribute the current to different points in the city, where its E. M. F. is reduced by ordinary transformers, from 1100 volts to 50 or 100, as may be required by the different lamp circuits; each dynamo being capable of feeding 1250 incandescent lamps, of 16 candle-power, a total of 8750.

In each dynamo circuit at the power station is a potential indicator, connected with a small transformer in which the E. M. F. of a shunt current is reduced in the ratio of 30 to 1. The secondary current of this transformer traverses also the secondary coil of another transformer, called a compensator, the primary coil of which is traversed by the main current, and the connection is so made that the compensator introduces into the indicator circuit a counter E. M. F. equal to the fall of potential in the main circuit, so that the indicator shows the E. M. F. at the Portland end of the line; and any variation from a constant E. M. F. of 3300 volts is corrected by varying the strength of the field current of the dynamo. The fall of potential in each feeder circuit connected with the main transformers at the substation is indicated in the same manner, and the E. M. F. adjusted by a regulator consisting of a transformer, the primary coil of which is on a shunt between the feeder mains, while the secondary coil is on the main feeder circuit, and is divided into sections, each of which has an E. M. F. equal to 1 per cent of that in the primary coil; and any number of these sections may be connected in series with the feeder circuit by a switch, and the E. M. F. varied as required. The method of regulation at each station, as described, is illustrated by the diagram in Fig. 95.

This was the first waterpower station in the United States, equipped for electric lighting by the alternating current, with long distance transmission. It has been in successful operation for three years with satisfactory results, both as to the working of the apparatus and the cost of maintenance; the operation of the dynamos being described as admirable, and the transformers not having cost a cent for repairs.

Telluride Waterpower Station.—Near Telluride, Colorado, is a waterpower station from which power is electrically transmitted to the Gold King mill, nearly three miles distant, where it is employed for operating crushers and stamps. It was equipped when first constructed with a Westinghouse alternating current dynamo of 100 horse-power, operated by a Pelton turbine wheel, driven by water received through a steel pipe two feet in diameter, under a head of 320 feet. The general construction of this dynamo is the same as that of the dynamos employed at the Willamette Falls station, but its field winding is composite, part of the magnets being excited by the armature current of a separate, direct current machine, and the others by a current from its own armature, which is made direct by an apparatus equivalent to a two-segment commutator; the adjustment being such that the E. M. F. of the current delivered through the mains rises as the current strength increases, compensating for the fall of potential in the line, and keeping the E. M. F. at the motor constant at 3000 volts. The speed is 833 revolutions per minute, producing 10000 alternations of current, and the switch-board connections and apparatus, including that for indicating the fall of potential in the line, are the same as those at Willamette Falls, as shown in Fig. 96.

The main current flows directly to the motor, at the mill, without transformation; the only transformers employed being the small ones connected with the indicators on the shunt circuits. The motor is the same in size, horse-power, and general construction as the dynamo, and runs in synchronism with it; but is excited by a current from its own armature, obtained from a special winding parallel with the main armature coils, and connected with the field coils by a circuit in which the current is made direct by a commutator. A small Tesla motor, of special construction, is employed as a starter for the large motor, and is connected with the mains by a parallel circuit, as shown. The armatures of both motors are belted to a countershaft, on which the ratio of

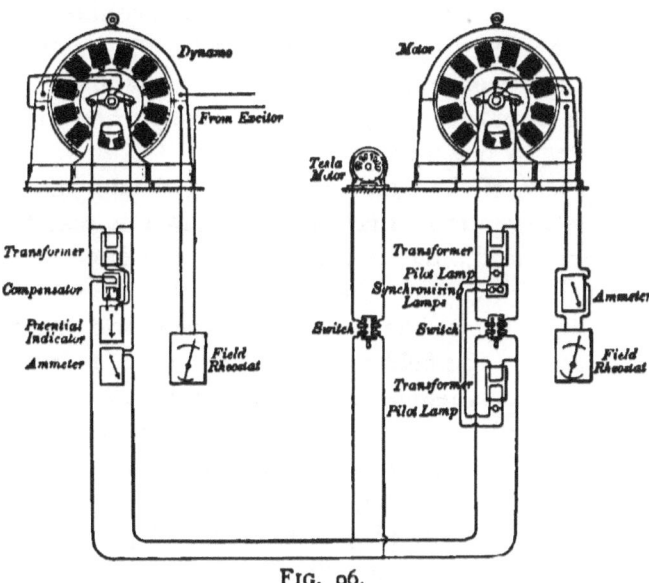

FIG. 96.

size between the pulleys is such as to give the armature of the large motor a little higher speed than that of the small one. When the circuit of the small motor is closed its armature quickly attains its normal speed, putting the armature of the large one in rotation, at a speed somewhat higher than that of the dynamo, and causing it to generate a self-exciting current, at the normal E. M. F. of the circuit. The small motor is then switched off, and the speed of the large one gradually decreases till it is approximately equal to that of the dynamo ; the relative speed of each machine being indicated by the degree of illumination in incandescent lamps connected in series with the secondary coils of two transformers, whose primary coils are connected respectively with the circuit of each machine, as shown; the illumination decreasing, from decrease of current, as the speeds of the two machines approach equality. When the proper relative speed, as thus indicated, is attained, the main circuit of the large motor is closed by its switch, and it is connected with the mill machinery by its friction clutch; the small motor having been disconnected by its clutch and brought to rest. The whole operation of starting is accomplished in about two minutes by one man.

If the speed of the motor, on starting, should happen to be a little lower than that of the dynamo, it may rise to the proper speed; but if much lower, it will continue to decrease, in which case the switch of the large motor is opened, and that of the small one closed, and the speed thus restored. The field current of the motor, as indicated by an ammeter, is regulated, on starting, by a rheostat, and requires no further adjustment for the varying loads.

The line runs across a rough country, ascending a mountain, at the power station, to a hight of 2500 feet, at an angle, in some places, of 45 degrees; and parts of it are practically inaccessible in winter, the snow being sometimes on a level with the tops of the poles. Special protection is required against lightning, to which this region is peculiarly liable; 40 discharges through the lightning arresters, having, on one occasion, occurred in 40 minutes.

The successful operation of the plant, under these unfavorable line conditions, and with a comparatively new type of electric apparatus, since its completion in June 1891, has inspired such confidence that extensive additions have been made, both for power and lighting; which indicates, that for the former purpose as well as the latter, the employment of the alternating current, with long distance transmission, is passing from the experimental to the practical stage.

Direct Connected Dynamos.—The direct connection of the armature of the dynamo with the shaft of the steam engine, without the intervention of belting, has certain advantages which make it desirable where such connection is practicable. As every intervening connection involves a certain percentage of loss in the transmission of power from the engine to the dynamo, it is evident that direct connection gives the maximum economy in transmission, besides the saving in first cost, repair, and renewal of belts and countershafting, and interest on capital thus invested; a considerable item in large stations. By this means, not only is the loss by friction and inertia of belts and countershafting eliminated but also the loss and irregularity due to the slipping of the belt.

But this connection is practicable only when the relative speed of engine and dynamo are properly equalized; which can be done with the latest improved triple expansion engines and multipolar dynamos. Several of the new Edison central stations have been equipped in this manner; two dynamos being directly connected with each engine, one on each side. But this system has not been generally adopted, and its adoption in new stations will, of course, depend largely on its practical success and advantages, as compared with the old system, as shown in the stations where it has been introduced.

INDEX.

A

	PAGE
Adjustable crossing, the Ramsay	153, 154
Alternating current motors	65–80
" " " , single phase	78–80
" " motor, the Brown single phase	79, 80
" " " , " Stanley-Kelly	72–78
" " " , " Tesla	66–72
Ampere, the	5
Applications of the stationary motor	81–136
Armature, the	8–10
Atkinson switch, the	152, 153

B

Boston trolley, the	143, 144
Brown single phase alternating current motor, the	79, 80
Brushes, the	10, 11
Brushes, position of the	17, 18
Burton electric heater, the	183–186

C

Central station construction and equipment	208–233
" " No. 1, Chicago Edison	209–215
" " " 5, " "	215–221
" " , Cicero and Proviso street railway	221–225
" " , development of the	208, 209

INDEX.

	PAGE
C. & C. small motors, the	53–56
" " " standard motor, the	52, 53
C. G. S. units	4
Chicago Edison central station No. 1	209–215
" " " " " 5	215–221
Cicero and Proviso street railway central station	221–225
City and South London underground electric railway, the	198–200
Clamps, insulators and	148–150
Classification of motors	7, 8
Closed conduits	192–195
Coal cutter, the New Arc electric	133–136
" ", " Sperry pick	132, 133
Commercial measurement of electric energy	113, 114
Commutation	15, 16
Commutator, the	10
Compression spring trolley, the	146, 147
Conduits, closed	192–195
Conduit railway, the Love	189–192
" ", " Siemens-Halske	187–189
" ", " Wheless	192–195
" system, the	186–195
Connections, controller	166–168
", rheostat	30–32
Conservation of energy, the	1, 2
Constant current and constant potential motors	24–27
Construction, coreless	17
" and equipment, central station	208–233
", line	137–140
" of direct current motors	8–12
Contact, the Siemens-Halske sliding	147, 148
" , " tube and piston	148
Controller	165, 166
" connections	166–168
Copper wire table	6
Coreless construction	17
Counter electromotive force	19–21
Cranes, electric traveling	108, 109
Crossing, right-angled	153
" , the Ramsay adjustable	153, 154
Current	3

INDEX. 237

	PAGE
Current reversal, effect of, on rotation	18, 19
Currents, eddy	21, 22
Curtis single reduction motor, the	173–176

D

Definitions	1–5
Dental apparatus, electric operation of	114–116
Designing, motor	34–41
Detroit motor, the	56–59
Development of the central station	208, 209
Diamond drill, the electric	128–130
Direct connected dynamos	233
" current motors, construction of	8–12
" " motor, operation of the	18
Disconnector, the Johnston	155–157
Dock hoists, electric	104–108
Double trolley, the	148
Drills, electric operation of ship	117
Drill, the Edison electric percussion	117–123
", " electric diamond	128–130
", " Van Depoele electric percussion	123–128
Dynamos, direct connected	233
Dynamo, operation of the motor as a	14, 15
Dyne, the	4

E

Eddy currents	21, 22
" motor, the	59–61
Edison electric percussion drill, the	117–123
" small motor, the	48–52
" standard motor, the	47, 48
Effect of current reversal on rotation	18, 19
Electric coal cutter, the New Arc	133–136
" " ", " Sperry pick	132, 133
" diamond drill, the	128–130
" dock hoists	104–108
" elevators	88–104
" elevator, the Otis	89–96
" " , " Sprague-Pratt	96–104

	PAGE
Electric energy, commercial measurement of	113, 114
" " , sources of	7
" fans and ventilators	84
" haulage in mills and factories	206, 207
" " " mines	204–206
" heat and mechanical energy	32–34
" heater, the Burton	183–186
" heating of cars	181–186
" horse-power	5
" lighting of cars	180, 181
" mining apparatus, various	136
" motor, principles of the	7–41
" operation of dental apparatus	114–116
" " " medical and surgical apparatus	116, 117
" " " pipe organs	84–88
" " " printing presses	109–113
" " " ship drills	117
" percussion drill, the Edison	117–123
" " ", " Van Depoele	123–128
" pump, the triplex	130–132
" railways and railway motors	137–207
" " , elevated and underground	195–200
" railway, the City and South London underground	198–200
" " , " Liverpool elevated	196–198
" terms	2–4
" traveling cranes	108, 109
" units	4, 5
Electromotive force	2, 3
" " , counter	19–21
Elevators, electric	88–104
Elevator, the Otis electric	89–96
" , " Sprague-Pratt electric	96–104
Elevated and underground electric railways	195–200
" electric railway, the Liverpool	196–198
Emmet trolley, the	144–146
" switch, the	152
Energy, commercial measurement of electric	113, 114
" , electric heat and mechanical	32–34
" , sources of electric	7
" , the conservation of	1, 2

INDEX. 239

	PAGE
Erg, the	4
Excelsior motor, the	42–47
Experiment, the Frankfort-Lauffen	226, 227

F

Fans and ventilators, electric	84
Feeders	140
Field-magnets, the	11, 12
Force, counter electromotive	19–21
" , electromotive	2, 3
Frankfort-Lauffen experiment, the	226, 227

G

Gearless motors	176–180
" motor, the Short	176–180
General remarks on electric railways and railway motors	137
" " " motor applications	81–84

H

Haulage in mills and factories, electric	206, 207
" " mines, electric	204–206
Heat and mechanical energy, electric	32–34
Heater, the Burton electric	183–186
Heating of cars, electric	181–186
Hoists, electric dock	104–108
Horse-power, the electric	5

I

Induction	3, 4
Insulators and clamps	148–150

J

Johnston disconnector, the	155–157

L

Law, Ohm's	3
Line-breaker, trolley	154, 155

	PAGE
Line construction	137–140
Lighting of cars, electric	180, 181
Lightning-arrester	168–170
Liverpool elevated electric railway, the	196–198
Long distance transmission, water power stations and	225–233
Loss of power	22
Love conduit railway, the	189–192

M

Measurement of electric energy, commercial	113, 114
Mechanical energy, electric heat and	32–34
Medical and surgical apparatus, electric operation of	116, 117
Mills and factories, electric haulage in	206, 207
Mining apparatus, various electric	136
Mines, electric haulage in	204–206
Motor, applications of the stationary	81–136
" designing	34–41
" , operation of the, as a dynamo	14, 15
" , " " " direct current	18
" , principles of the electric	7–41
" , the, a means of applying power	7
" , " Brown single phase alternating current	79, 80
" , " C. & C. standard	52, 53
" , " Curtis single reduction	173–176
" , " Detroit	56–59
" , " Eddy	59–61
" , " Edison small	48–52
" , " " standard	47, 48
" , " Excelsior	42–47
" , " Perret	61–65
" , " Series wound	12
" , " Short gearless	176–180
" , " shunt wound	12–14
" , " Stanley-Kelly alternating current	72–78
" , " Tesla alternating current	66–72
" , " Thomson-Houston waterproof single reduction	170–173
" , " Westinghouse single reduction	161–168
Motors, alternating current	65–80
" , classification of	7, 8

INDEX. 241

	PAGE
Motors, compared, series, and shunt	22–24
", constant current and constant potential	24–27
", construction of direct current	8–12
", electric railways and railway	137–207
", gearless	176–180
", railway	157–180
", single phase alternating current	78–80
", stationary	42–80
", the C. & C. small	53–56

N

Neutral line, polarity and	16, 17
" ", position of the poles and	21
New Arc electric coal cutter, the	133–136

O

Ohm's law	3
Ohm, the	4
Otis electric elevator, the	89–96
Operation of dental apparatus, electric	114–116
" " medical and surgical apparatus, electric	116, 117
" " pipe organs, electric	84–88
" " printing presses, electric	109–113
" " the direct current motor	18
" " " motor as a dynamo	14, 15
" " ship drills, electric	117

P

Percussion drill, the Edison electric	117–123
" " " Van Depoele electric	123–128
Perret motor, the	61–65
Pipe organs, electric operation of	84–88
Polar rotation	19
Polarity and neutral line	16, 17
Poles	140
Position of the brushes	17, 18
" " " poles and neutral line	21

242 INDEX.

	PAGE
Potential	2
Power, loss of	22
", the motor a means of applying	7
Principles of the electric motor	7–41
Printing presses, electric operation of.	109–113
Pump, the triplex electric	130–132

R

Railways and railway motors, electric	137–207
", elevated and underground	195–200
Railway motors	157–180
", the City and South London underground electric	198–200
", " Liverpool elevated electric	196–198
", " Love conduit	189–192
", " Siemens-Halske conduit	187–189
", " Wheless conduit	192–195
Ramsay adjustable crossing, the	153, 154
Remarks, general, on electric railways and railway motors	137
", ", " motor applications	81–84
Resistance	3
Reversal of rotation	17
Rheostat connections	30–32
Rheostat, the	27–30
Right-angled crossing	153
Rotation, effect of current reversal on	18, 19
", polar	19
", reversal of	17

S

Series and shunt motors compared	22–24
" wound motor, the	12
Short gearless motor, the	176–180
Ship drills, electric operation of	117
Shunt wound motor, the	12–14
Siemens-Halske conduit railway, the	187–189
" " sliding contact, the	147, 148
Single phase alternating current motors	78–80
" " " " motor, the Brown	79, 80

INDEX. 243

	PAGE
Single reduction motor, the Curtis	173–176
" " " , " Thomson-Houston waterproof	170–173
" " " , " Westinghouse	161–168
Sliding contact, the Siemens-Halske	147, 148
Small motor, the Edison	48–52
" motors, the C. & C.	53–56
Sources of electric energy	7
Sperry pick electric coal cutter, the	132, 133
Sprague-Pratt electric elevator, the	96–104
Standard motor, the C. & C.	52, 53
" " , " Edison	47, 48
Stanley-Kelly alternating current motor, the	72–78
Stationary motors	42–80
" motor, applications of the	81–136
Storage battery traction	200–204
Street railway central station, Cicero and Proviso	221–225
Switches	151–153
Switch, the Atkinson	152, 153
" , " Emmet	152
" , " three-way	151, 152
System, the conduit	186–195

T

Table, copper wire	6
Telluride water-power station	230–233
Terms, electric	2–4
Tesla alternating current motor, the	66–72
Thomson-Houston waterproof single reduction motor, the	170–173
Three-way switch, the	151, 152
Traction, storage battery	200–204
Transmission, waterpower stations and long distance	225–233
Traveling cranes, electric	108, 109
Triplex electric pump, the	130–132
Trolleys	140–147
Trolley line-breaker	154, 155
" , the Boston	143, 144
" , " compression spring	146, 147
" , " double	148
" , " Emmet	144–146
Tube and piston contact, the	148

U

	PAGE
Units, C. G. S.	4
" , electric.	4, 5

V

Van Depoele electric percussion drill, the	123–128
Various electric mining apparatus	136
Ventilators, electric fans and	84
Volt, the	4

W

Waterpower stations and long distance transmission	225–233
" station, Telluride	230–233
" " , Willamette falls	227–230
Waterproof single reduction motor, the Thomson-Houston	170–173
Watt, the	5
Westinghouse single reduction motor, the	161–168
Wheless conduit railway, the	192–195
Willamette falls water power station	227–230
Wire table, copper	6

www.ingramcontent.com/pod-product-compliance
Lightning Source LLC
Chambersburg PA
CBHW031728230426
43669CB00007B/281